HANDBOOK OF RESEARCH ON FOOD SCIENCE AND TECHNOLOGY

Volume 3

Functional Foods and Nutraceuticals

HANDBOOK OF RESEARCH ON FOOD SCIENCE AND TECHNOLOGY

Volume 3

Functional Foods and Nutraceuticals

Edited by

Mónica Lizeth Chávez-González, PhD

José Juan Buenrostro-Figueroa, PhD

Cristóbal N. Aguilar, PhD

Apple Academic Press Inc.
3333 Mistwell Crescent
Oakville, ON L6L 0A2 Canada

Apple Academic Press Inc.
9 Spinnaker Way
Waretown, NJ 08758 USA

First issued in paperback 2021

Exclusive worldwide distribution by CRC Press, a member of Taylor & Francis Group

Handbook of Research on Food Science and Technology, Volume 3: Functional Foods and Nutraceuticals

ISBN 13: 978-1-77463-530-8 (pbk)
ISBN 13: 978-1-77188-720-5 (hbk)

Handbook of Research on Food Science and Technology, 3-volume set

ISBN 13: 978-1-77188-721-2 (hbk)

Library and Archives Canada Cataloguing in Publication

Handbook of research on food science and technology / edited by Mónica Lizeth Chávez-González, PhD, José Juan Buenrostro-Figueroa, PhD, Cristóbal N. Aguilar, PhD.

Includes bibliographical references and indexes.
Contents: Volume 3. Functional foods and nutraceuticals.
Issued in print and electronic formats.
ISBN 978-1-77188-720-5 (v. 3 : hardcover).--ISBN 978-0-429-48782-8 (v. 3 : PDF)

1. Food industry and trade--Technological innovations. 2. Food--Research. 3. Food--Biotechnology. 4. Food--Composition. 5. Functional foods. I. Chávez-González, Mónica Lizeth, 1987-, editor II. Buenrostro-Figueroa, José Juan, 1985-, editor III. Aguilar, Cristóbal Noé, editor

TP370.H36 2018 664'.024 C2018-906012-3 C2018-906013-1

CIP data on file with US Library of Congress

Apple Academic Press also publishes its books in a variety of electronic formats. Some content that appears in print may not be available in electronic format. For information about Apple Academic Press products, visit our website at **www.appleacademicpress.com** and the CRC Press website at **www.crcpress.com**

CONTENTS

ABOUT THE EDITORS

Mónica Lizeth Chávez-González, PhD
Full Professor, School of Chemistry of the Universidad Autónoma de Coahuila, Mexico

Mónica Lizeth Chávez-González, PhD, is a Full Professor at the School of Chemistry of the Universidad Autónoma de Coahuila, Mexico, where she develops her work in the Food Research Department. Dr. Chávez-González's experience is in the areas of fermentation processes, microbial biotransformation, enzyme production, valorization of food industrial wastes, extraction of bioactive compounds, and chemical characterization. She is a member of the Sociedad Mexicana de Biotecnología y Bioingeniería and the Asociación Mexicana para la Protección a los Alimentos affiliate of the International Association for Food Protection. She was awarded with the "Juan Antonio de la Fuente" medal for academic excellence and the "Ocelotl" prize for best tecnhological innovation purpose, both given by the Universidad Autónoma de Coahuila. She earned her PhD in Food Science and Technology with an emphasis on valorization of food industrial waste under the tutelage of Dr. Cristóbal N. Aguilar.

José Juan Buenrostro-Figueroa, PhD
Researcher, Research Center for Food and Development, A.C., Mexico

José Juan Buenrostro-Figueroa, PhD, is a Researcher at the Research Center for Food and Development, A.C., Mexico. Dr. Buenrostro has experience in bioprocess development, including microbial processes for enzyme production and recovery of bioactive compounds; valorization of agroindustrial by-products, and extraction and characterization of bioactive compounds from plants and agroindustrial wastes. He has published 17 papers in indexed journals, five book chapters, four patent requests, and more than 45 contributions at scientific meetings. Dr. Buenrostro has been a member of S.N.I. (National System of Researchers, Mexico), the Mexican Society of Biotechnology and Bioengineering (SMBB), and the Mexican Society for Food Protection affiliate of the International Association for Food Protection.

He became a Food Engineer at the Antonio Narro Agrarian Autonomous University. He earned his MSc and PhD degrees in Food Science and Technology from the Autonomous University of Coahuila, México, where he worked on the development of bioprocesses for the valorization of agroindustrial by products. He also worked at in the Biotechnology Department of the Metropolitan Autonomous University, Mexico City, Mexico.

Cristóbal N. Aguilar, PhD

Full Professor and Dean, School of Chemistry, Universidad Autónoma de Coahuila, Mexico

Cristóbal N. Aguilar, PhD, is a Full Professor and Dean of the School of Chemistry at the Universidad Autónoma de Coahuila, Mexico. Dr. Aguilar has published more than 160 papers published in indexed journals, more than 40 articles in Mexican journals, as well as 16 book chapters, eight Mexican books, four international books, 34 proceedings, and more than 250 contributions in scientific meetings. Professor Aguilar is a member of the National System of Researchers of Mexico (SNI) and has received several prizes and awards, the most important are the National Prize of Research (2010) of the Mexican Academy of Sciences, the "Carlos Casas Campillo 2008" prize of the Mexican Society of Biotechnology and Bioengineering, the National Prize AgroBio-2005, and the Mexican Prize in Food Science and Technology from CONACYT-Coca Cola México in 2003. He is also a member of the Mexican Academy of Science, the International Bioprocessing Association (IFIBiop), and several other scientific societies and associations. Dr. Aguilar has developed more than 21 research projects, including six international exchange projects. He has been advisor of 18 PhD theses, 25 MSc theses, and 50 BSc theses.

He became a Chemist at the Universidad Autónoma de Coahuila, Mexico, and earned his MSc degree in Food Science and Biotechnology at the Autonomous University of Chihuahua, México. His PhD degree in Fermentation Biotechnology was awarded by the Autonomous University of Metropolitana, Mexico. Dr. Aguilar also performed postdoctoral work at the Department of Biotechnology and Molecular Microbiology at Research Institute for Development (IRD) in Marseille, France.

CONTRIBUTORS

Cristóbal N. Aguilar
Bioprocesses Group, Food Research Department, Faculty of Chemistry,
Autonomous University of Coahuila, Saltillo, 25280, Coahuila, Mexico,
E-mail: cristobal.aguilar@uadec.edu.mx

Miguel A. Aguilar-González
Cinvestav-IPN. Unit Saltillo, Coahuila, Mexico

Pedro Aguilar-Zárate
Department of Engineering, Tecnológico Nacional de México, Campus Ciudad Valles, 79010,
Ciudad Valles, San Luis Potosí, Mexico, Tel.: +52 844 6087976,
E-mail: pedro.aguilar@tecvalles.mx

Jorge Alejandro Aguirre-Joya
School of Health Science, Universidad Autonoma de Coahuila, North Unit, Piedras Negras,
Coahuila, Mexico, E-mail: jorge_aguirre@uadec.edu.mx

Olga B. Alvarez-Perez
Group of Bioprocesses and Bioproducts, Food Research Department, School of Chemistry,
Universidad Autonoma de Coahuila, 25280 Saltillo, Coahuila, México

Dulce G. Argüello-Esparza
Biorefinery Group, Food Research Department, Faculty of Chemical Sciences,
Autonomous University of Coahuila, 25280, Saltillo, Coahuila, Mexico,
Tel.: (+52) 844 416 12 38

Juan Alberto Ascacio-Valdés
Food Research Department, Faculty of Chemistry, Autonomous University of Coahuila,
Blvd. Venustiano Carranza, Col. República Oriente, 25280, Saltillo, Coahuila, Mexico,
Tel.: 8442450073

Victor Daniel Boone-Villa
School of Medicine, Universidad Autonoma de Coahuila, North Unit, Piedras Negras,
Coahuila, Mexico

José Del Bosque-Moreno
School of Health Science, Universidad Autonoma de Coahuila, North Unit, Piedras Negras,
Coahuila, Mexico

María L. Carrillo-Inungaray
Food Research Laboratory, Autonomous University of San Luis Potosí, 79060, Ciudad Valles,
San Luis Potosí, México

Luis Enrique Cobos-Puc
Department of Bioactive Compounds, Faculty of Chemical Sciences,
Autonomous University of Coahuila, 25280 Saltillo, Coahuila, Mexico

Juan C. Contreras-Esquivel
Food Research Department, School of Chemistry, Autonomous University of Coahuila, Boulevard Venustiano Carranza and José Cárdenas s/n, República Oriente, Saltillo 25280, Coahuila, México

Sarai Escobedo-García
Food Research Department, School of Chemistry, Autonomous University of Coahuila, Boulevard Venustiano Carranza and José Cárdenas s/n, República Oriente, Saltillo 25280, Coahuila, México

Adriana C. Flores-Gallegos
Food Research Department, School of Chemistry, Autonomous University of Coahuila, Boulevard Venustiano Carranza and José Cárdenas s/n, República Oriente, Saltillo 25280, Coahuila, México

Melissa Flores-García
Autonomous University of Coahuila, Nanobioscience Group, School of Chemistry, Blvd. V. Carranza and José Cárdenas Valdés s/n Col. Republic East, ZIP 25280, Saltillo, Coahuila, Mexico, Tel.: +52–844–416–92–13

Dulce A. Flores-Maltos
Institute for Biotechnology and Bioengineering, Centre of Biological Engineering, University of Minho, 4710–057 Braga, Portugal, E-mail: abril.maltos@ceb.uminho.pt

Mayela Govea-Salas
Autonomous University of Coahuila, Nanobioscience Group, School of Chemistry, Blvd. V. Carranza and José Cárdenas Valdés s/n Col. Republic East, ZIP 25280, Saltillo, Coahuila, Mexico, Tel.: +52-844-416-92-13

Anna Iliná
Food Research Department, Faculty of Chemistry, Autonomous University of Coahuila, Blvd. Venustiano Carranza, Col. República Oriente, 25280, Saltillo, Coahuila, Mexico, Tel.: 8442450073, +52-844-416-92-13,
E-mails: annailina@uadec.edu.mx; anna_ilina@hotmail.com

Miguel Ángel De León-Zapata
Group of Bioprocesses and Bioproducts, Food Research Department, School of Chemistry, Universidad Autonoma de Coahuila, 25280 Saltillo, Coahuila, México

Lluvia Itzel López López
Food Research Department, Faculty of Chemistry, Autonomous University of Coahuila, Blvd. Venustiano Carranza, Col. República Oriente, 25280, Saltillo, Coahuila, Mexico, Tel.: 8442450073

Ricardo Guadalupe López-Ramos
Department of Bioactive Compounds, Faculty of Chemical Sciences, Autonomous University of Coahuila, 25280 Saltillo, Coahuila, Mexico

Jesús A. Martínez-Salas
Food Research Laboratory, Autonomous University of San Luis Potosí, 79060, Ciudad Valles, San Luis Potosí, México

Mariela Michel
Food Research Department, School of Chemistry, Autonomous University of Coahuila, 25280, Saltillo, Coahuila, Mexico

Diana B. Muñiz-Márquez
Department of Engineering, Tecnológico Nacional de México, Campus Ciudad Valles, 79010, Ciudad Valles, San Luis Potosí, Mexico, Tel.: +52 844 6087976

Nestor Humberto Obregón-Sánchez
School of Health Science, Universidad Autonoma de Coahuila, North Unit, Piedras Negras, Coahuila, Mexico

Alma Rosa Paredes-Ramírez
Autonomous University of Coahuila, Laboratory Animal Center, School of Medicine. Francisco Murguía South No. 205. ZIP 25000, Saltillo, Coahuila, Mexico

Brian Picazo
Bioprocesses Group, Food Research Department, Faculty of Chemistry, Autonomous University of Coahuila, Saltillo, 25280, Coahuila, Mexico

Rodolfo Ramos-González
Autonomous University of Coahuila, Blvd. V. Carranza and José Cárdenas Valdés s/n Col. Republic East, ZIP 25280, Saltillo, Coahuila, Mexico

Diana Jasso de Rodríguez
Department of Plant Breeding, Antonio Narro Autonomous Agrarian University, Calzada Antonio Narro # 1923, Colonia Buenavista, 25315, Saltillo, Coahuila, Mexico

Raúl Rodríguez-Herrera
Food Research Department, School of Chemistry, Autonomous University of Coahuila, Boulevard Venustiano Carranza and José Cárdenas s/n, República Oriente, Saltillo 25280, Coahuila, México

Rosa M. Rodríguez-Jasso
Biorefinery Group, Food Research Department, Faculty of Chemical Sciences, Autonomous University of Coahuila, 25280, Saltillo, Coahuila, Mexico, Cluster of Bioalcohols, Mexican Centre for Innovation in Bioenergy (Cemie-Bio), Mexico, Tel.: (+52) 844 416 12 38, E-mail: rrodriguezjasso@uadec.edu.mx

Romeo Rojas Molina
Group of Bioprocesses and Bioproducts, Food Research Department, School of Chemistry, Universidad Autonoma de Coahuila, 25280 Saltillo, Coahuila, México

Héctor A. Ruiz
Biorefinery Group, Food Research Department, Faculty of Chemical Sciences, Autonomous University of Coahuila, 25280, Saltillo, Coahuila, Mexico, Cluster of Bioalcohols, Mexican Centre for Innovation in Bioenergy (Cemie-Bio), Mexico, Tel.: (+52) 844 416 12 38

Jesús A. Salas-Tovar
Food Research Department, School of Chemistry, Autonomous University of Coahuila, Boulevard Venustiano Carranza and José Cárdenas s/n, República Oriente, Saltillo 25280, Coahuila, México

Elda Patricia Segura-Ceniceros
Autonomous University of Coahuila, Nanobioscience Group, School of Chemistry, Blvd. V. Carranza and José Cárdenas Valdés s/n Col. Republic East, ZIP 25280, Saltillo, Coahuila, Mexico, Tel.: +52-844-416-92-13

Crystel Aleyvick Sierra-Rivera
Department of Bioactive Compounds, Faculty of Chemical Sciences,
Autonomous University of Coahuila, 25280 Saltillo, Coahuila, Mexico

Sonia Yesenia Silva-Belmares
Food Research Department, Faculty of Chemistry, Autonomous University of Coahuila,
Blvd. Venustiano Carranza, Col. República Oriente, 25280, Saltillo, Coahuila, Mexico,
Tel.: 8442450073, E-mail: yesenia_silva@hotmail.com and yesenia_silva@uadec.edu.mx

Bartolomeu W. S. Souza
Department of Fisheries Engineering, Federal University of Ceará,
Department of Fisheries Engineering, Campus do Pici S / N, Block 827,
Zip Code: 60455-970, Fortaleza-Ceará, Brazil, E-mail: souzabw@gmail.com

José A. Teixeira
Institute for Biotechnology and Bioengineering, Centre of Biological Engineering,
University of Minho, 4710–057 Braga, Portugal

Monica Valencia
Food Research Department, Faculty of Chemistry, Autonomous University of Coahuila,
Blvd. Venustiano Carranza, Col. East Republic, 25280, Saltillo, Coahuila, Mexico

Dalia Vásquez-Bahena
Healthcare Business and Computer Technology S.A. de C.V. Tlaxcala No. 146/705,
Col. Roma South ZIP. 06760, Cuauhtémoc, Mexico City, Mexico

Janeth Ventura-Sobrevilla
Group of Bioprocesses and Bioproducts, Food Research Department, School of Chemistry,
Universidad Autonoma de Coahuila, 25280 Saltillo, Coahuila, México

Perla Yaneth Villa Silva
Food Research Department, Faculty of Chemistry, Autonomous University of Coahuila,
Blvd. Venustiano Carranza, Col. República Oriente, 25280, Saltillo, Coahuila, Mexico,
Tel.: 8442450073

Jorge E. Wong-Paz
Department of Engineering, Tecnológico Nacional de México, Campus Ciudad Valles, 79010,
Ciudad Valles, San Luis Potosí, Mexico, Tel.: +52 844 6087976

ABBREVIATIONS

AAPH	2,2-azobis-(2-amidinopropane) dihydrochloride
ACE	angiotensin I-converting enzyme
AD	Alzheimer's disease
ALA	alpha linolenic acid
BCS	breast-conserving surgery
CAM	Crassulacean acid metabolism
CN	caseins
CVD	cardiovascular disease
DHA	docosahexaenoic acid
DM	diabetes mellitus
DNA	deoxyribonucleic acid
DP	degree of polymerization
DPPH	2,2-diphenyl-1-picrylhydrazyl
EDUF	electrodialysis with ultrafiltration membranes
EGCG	epigallocatechin–3-gallate
EPA	eicosapentaenoic acid
ESI	electrospray ionization
FAO	Food and Agriculture Organization
FASN	fatty acid synthase
FOS	Fructooligosaccharides
FOSHU	foods for specific health use
FPH	fish protein hydrolysates
FUFOSE	European Commission Concerted Action on Functional Food Science in Europe
GABA	γ-amino-butyric acid
GALT	gut associated lymphoid system
GOS	Galactooligosaccharides
HBV	hepatitis B viruses
HCC	Hepatocellular carcinoma

HCV	hepatitis C viruses
HDL	high density lipoproteins
HPLC	high performance liquid chromatography
HSOs	Homeostatic® soil organisms
IEC	ion-exchange chromatography
IEF	isoelectric focusing
Ig	immunoglobulins
IMO	isomaltulose
LF	lactoferrin
MALDI	matrix assisted laser desorption ionizations
MALDI-TOF	matrix assisted laser desorption-ionization time-of-flight
MDA	malondialdehyde
MIR	mid-infrared spectroscopy
MS	mass spectrometry
MSSA	methicillin-sensitive *S. aureus*
MUFA	monounsaturated fatty acids
NCD	non-communicable diseases
NIRS	near infrared spectroscopy
NMR	nuclear magnetic resonance
NO	nitric oxide
NO-cGMP	nitric oxide cyclic guanosine monophosphate
NRPS	nonribosomal peptide synthesis
NSCLC	non-small cell lung cancer
PD	Parkinson's disease
PRPs	proline-rich proteins
PUFA	polyunsaturated fatty acids
PUFAs ω-3	omega-3 polyunsaturated fatty acids
RCTs	randomized clinical trials
RNS	reactive nitrogen species
ROS	reactive oxygen Species
SCFA	short chain fatty acids
SCLC	small cell lung cancer
SEC	size exclusion chromatography

SFA	saturated fatty acids
TCA	trichloroacetic acid
TFA	trans fatty acids
TLC	thin layer chromatography
TOF	time-of-flight
UF	ultrafiltration
UHPLC	ultra-high pressure liquid chromatography
WHO	World Health Organization
XOS	xylooligosaccharides
α-LA	α-lactalbumin
β-LG	β-lactoglobulin

PREFACE

Human population is growing dramatically, and it will probably reach 10 billion world inhabitants sooner than estimated. It implies important challenges never faced before by humans who need to organize and work on the development of mega-efforts to ensure universal access to health care, food, water, sanitization, energy, education, and housing. These challenges, natural or man-made, obligate the scientific community to proactively seek new breakthrough food and nutrition solutions to ensure global food sustainability and nutrition security in the future. To achieve this, innovative solutions need to be considered throughout the whole food chain, inclusive of food choices and dietary patterns, in order to make significant improvements in the food supply, nutritional, and health status. In the case of foods, innovations in food processing techniques can significantly contribute to meeting the needs of the future world population with respect to quality, quantity, and sustainability of food intake.

Those in academe and industry focused on food science and technology are constantly redefining their traditional forms for new ways to face the threats of the twenty-first century, which is marked by multiple unprecedented environmental challenges that could threaten human survival. The combined impact of climate change, energy and water shortages, environment pollutants, shifting global population demographics, food safety, and growing disease pandemics all place undue stress on the planet's food system, already in a sensitive balance with its ecosystem.

Any changes to the food supply inevitably impact food, nutrition, and health trends and policies, particularly pertaining to food production, agricultural practices, dietary patterns, nutrition, and health guidance and management. As a result, there is an urgent need to find alternative solutions to improve the efficiency and sustainability in the food supply chain by reducing food waste and enhancing nutritional qualities of foods through the addition of nutraceuticals to prepare functional foods and intelligent foods.

The Food Research Group of the School of Chemistry at Universidad Autonoma de Coahuila (DIA-UAdeC) celebrates 25 years of existence and hard work, a period in which it has undergone a tremendous transformation in order to provide solutions and new technological alternatives to the problems demanded by the region, the country, and some international elements. To achieve this, the group grew in the number of researchers, and therefore various lines of research are studied. Today the DIA-UAdeC is formed by research groups in Bioprocesses and Bioproducts, Biorefineries, Biocontrol, Natural Products, Molecular Biology and Ommic Sciences, Glic-Biotechnology, Nano-Bioscience, Edible Coatings, Films and Membrane Technology, Food Engineering, Emerging Processing Technology, Food Science, and Functional Foods.

Two important Mexican postgraduate programs in Food Science and Technology are offered to Mexican and foreign students to whom the National Council of Science and Technology of Mexico (CONACYT) offers scholarships to carry out their MSc or PhD programs.

The consolidation of national scientific cooperation has allowed the prolongation and substantially improvement of the generation and application of knowledge with the Autonomous Metropolitan University, the Autonomous University of Chihuahua, the Autonomous University of Tamaulipas, the Autonomous Agrarian University "Antonio Narro," the Autonomous University of Nuevo León, the University of Colima, the Technological Institute of Durango, the Technological Institute of Ciudad Valles, the Technological Institute of Monterrey, and the centers of research CIQA, CINVESTAV, CIMAV, CIAD, CICATA, CIATEJ, among others.

Strong linkages of international cooperation with institutions and research centers around the world have been established and are now generating important results in the framework of scientific and technological cooperative projects and programs. They highlight their research partnerships with the University of Minho (Portugal); the University of Vigo (Spain); the University of Georgia (USA); the University of Marseille (France); the University of Valle and the National University of Colombia; the National University Nacional de Rosario, the National University of Rio Cuarto and the National University of La Plata (Argentina); Kannur University (India); Federal University of Pernambuco (Brazil); University of Torino (Italy); Jacobs University (Germany); Gachon University (Korea); and other important world-quality research centers including INL (Portugal); IRD and IMBE (France); ICIDCA (Cuba), and the Jawaharlal Nehru Tropical Botanic Garden & Research Institute (India).

For this reason, the research group has organized itself to celebrate its 25th anniversary by publishing a book that reflects the scientific and technological contributions in the field of food Science and technology generated by scientists of the DIA-UAdeC and some of its collaborators.

This *Handbook of Research in Food Science and Technology* consists of three volumes; (i) Food Technology and Chemistry, (ii) Food Biotechnology and Microbiology, and (iii) Functional Foods and Nutraceuticals, all of which will highlight the current trends and knowledge regarding the most recent innovations, emerging technologies, and strategies based on food design on a sustainable level. The handbook includes relevant information on the modernization of food industries, emerging technologies, sustainable packaging, food bioprocesses, food fermentation, food microbiology, functional foods, nutraceuticals, natural products, nano- and micro-technology, healthy product composition, innovative processes/bioprocesses for utilization of by-products, development of novel preservation alternatives, extending the shelf life of fresh products, and alternative processes requiring less energy or water, among other topics.

CHAPTER 1

TRENDS IN FUNCTIONAL FOOD IN NON-COMMUNICABLE DISEASES

VICTOR DANIEL BOONE-VILLA,[1]
NESTOR HUMBERTO OBREGÓN-SÁNCHEZ,[2]
JOSÉ DEL BOSQUE-MORENO,[2] and
JORGE ALEJANDRO AGUIRRE-JOYA[2]

[1] *School of Medicine, Universidad Autonoma de Coahuila, North Unit, Piedras Negras, Coahuila, Mexico*

[2] *School of Health Science, Universidad Autonoma de Coahuila, North Unit, Piedras Negras, Coahuila, Mexico, E-mail: jorge_aguirre@uadec.edu.mx*

ABSTRACT

The relationship of food with health maintenance and disease prevention has been well known since ancient times. However, the increasing acceptance of the connection between diet and health presents new opportunities and perspectives of food ingredients over physiological and metabolic functions and health and disease process in consumers. Non-communicable diseases (NCD) are responsible for 38 million deaths each year, approximately 28 million occur in low and middle-income countries. Cardiovascular disease, cancer, and diabetes are the NCD that most death causes (17.5, 8.2, and 1.5 million people, respectively). Among the risk factors associated with NCD, the most important ones are smoking, physical inactivity, harmful consumption of alcohol and unhealthy diets. In recent years functional foods have been used to treat different illnesses, including NCD. Functional foods are expressly made to improve health in patients besides of the nutritive function in the organism. In the present chapter, the most relevant studies on functional foods in NCDs, their health-disease relationship, their components, improvements, uses and advantages are discussed.

1.1 INTRODUCTION

Functional food is defined as the food that provides preventive, curative, and or protective effect against one or more diseases, further its nutritional benefits [1]. However specific denomination of what is functional food depends on each country, their internal legislation and the time that the country has been legislating, one example of a country with the high preoccupation of functional foods in Japan with 30 years of expertise in Foods for Specific Health Use (FOSHU). According to Ashwell [2] and Pravst [3] functional food can be one or a combination of following: a natural unmodified food with provided health effects; a vegetal or any food where a specific component has been improved through special growing, breeding or biotechnology conditions; a food where a component has been added to provide health benefits; a food where, by technological or biotechnological techniques a component has been removed to provide health benefits that in another way they cannot be achieved; a food whereas a component has been replaced or other with favorable effects. No matter different definitions for functional foods, their main purpose is to improve human health.

Characteristics of modern life, particularly, eating habits and physical activity have favored an increase in overweight, obesity, diabetes, hypertension, and cardiovascular disease (CVD) due that alimentation has turned excessive and inadequate, characterized by high caloric diet, rich in alcohol, animal fat, cholesterol, salt, and refined sugars with a low ratio between calories and nutrients [4]. These illnesses are also known as non-communicable diseases (NCD) [5] that by definition are chronic diseases that are non-transmissible and non-infectious [6]. The costs generated direct (those related to medical intervention need for healthcare as rehabilitation, drugs, and hospitalization) and indirect (identified as the loss of productivity and the need to replace the workers affected by any of these the diseases) by NCD are considerable [4].

Data described in the Global Action Plan for The Prevention and Control of Non-Communicable Diseases 2013–2020 describes that the NCD's bigger killers in the world are: cardiovascular diseases, cancers, chronic respiratory diseases, and diabetes, with more than 36 million people dying each year due to these diseases; this represents 63% of global deaths. 14 millions of these deaths occur in people between 30 and 70 years old, most of these deaths (86%) take place in low and middle-income countries [7].

Environmental behavior plays an important role in the development of NCD due to the interaction between foods, nutrients, and human genetic make-up [4]. Among the most important environmental behavior that presents risk factors, the eating habits and physical activity are present.

Raised blood pressure, overweight/obesity, hyperglycemia, and hyperlipidemia are the four metabolic changes that directly increase the risk of NCD apparition [5]. According to the Global Burden of Disease reported study, 19% of global deaths are attributable to raised blood pressure followed by obesity [8].

Keys et al. [9] developed the first study where coronary heart diseases were directly associated with diet, giving a new point of view to food, especially to the Mediterranean diet, that is synonymous with healthy and delicious food. Mediterranean diet is composed mainly of olive oil, garlic, wine, almonds, vegetables, and fish [10]. Also, diets rich in high-fiber foods and low glucose and fat apportion have been associated with a reduced cardiovascular risk [11].

Today functional foods are a part, with nutraceuticals and beverages, of a $150 billion dollar market, with a growth of 10% each year [12]. One of the reasons is that healthy foods have fundamental market conditions like science, innovation, and resolution of health problems such as heart attacks [13]. By the way, ethical fundaments, convenience, and scientific justification must support elaboration and commercialization of functional foods.

Recent advances in the food industry have led to the creation and recommendation for the consumption of food with bioactive compounds, such as omega 3 and 6 and natural antioxidants as polyphenols, that represents most abundant natural antioxidants in human diets [14, 15]. Polyphenols play an important role as antioxidants, what it means they reduce oxidative phenomena implicated in the aging process and oxidative stress chain, bot important damage causes in the body related to NCD development [16, 17].

Obesity is one of the most dangerous epidemics of XXI century, only in the European Union 35% of people are overweight; 20% in the United States of America and 39% of adults around the world are overweight and 13% presents obesity [5, 18]. This is an important fact since obesity is a key risk factor that triggers others NCD like diabetes mellitus (DM), hypertension, and metabolic syndrome. Obesity is also related to the development of several types of cancer [19].

During obesity, mitochondrial dysfunction and impaired energy metabolism are central alterations. In a hypercaloric diet or when nutrient surplus overwhelms the mitochondrial system, these organelles get damaged and its functions are impaired, with the accumulation of incompletely oxidized lipidic compounds derived from the Krebs cycle with the accumulation of fat and oxidative stress generation [6]. Due to this, a balanced and adequate diet can help maintain the health of people and functional food can reduce complications, prevent, and avoid illnesses and damage, especially the development of NCD such as cancer, DM, and CVD.

1.2 FUNCTIONAL FOODS AND CANCER PREVENTION

Cancer is one of the major health problems and it is causing one of each eight deaths worldwide. Approximately 25% of Americans will have cancer in their lifetime. Treatment usually involves the expensive and often traumatic use of drugs, surgery, and irradiation. The 13.3 million new cases of cancer in 2010 were predicted to cost US$ 290 billion, but the total costs were expected to increase to US$ 458 billion in the year 2030 on the basis of World Economic Forum in 2011 [20].

Food plays a very important role in the development and prevention of this disease without forgetting the rest of the risk factors that contribute to its incidence, including the genetic ones [21]. There are a series of food that due to his physicochemical components help to protect against the previously NCD's mentioned fruits and vegetables can be emphasized. In addition, there are other groups of foods that consumed in excess can favor the appearance of cancer, particular, the food with high contents of carcinogens, such as heavy metals, pesticides, and aromatic hydrocarbons (present in smoked foods).

The term functional food was first introduced in Japan in the mid-1980s and refers to processed foods containing ingredients that provide specific health functions in addition to being nutritious. The Institute of Medicine's Food and Nutrition Board/National Academy of Science [22] defined functional foods as "any food or food ingredient that may provide a health benefit beyond the traditional nutrients it contains." The European Commission Concerted Action on Functional Food Science in Europe (FUFOSE) regards a food as functional if it is satisfactorily demonstrated to modify beneficially one or more target functions of the body, in a way that it is relevant to an improved state of health and well-being and/or a reduction of disease risk, beyond adequate nutritional effects.

Dietary habits are linked with the development of cancer; diet with low whole grains is regarded as the main factor for the development of various cancers. The importance of food in cancer prevention has been documented. Food invariably is the source of various nutrients; however, certain foods go beyond their basic function of providing nutrients and possess health-enhancing properties. There have been numerous evidence-based studies about the effectiveness of functional food's cancer therapeutic and/or -preventive potential. Dietary factors causally linked with various cancers are high calorie fatty foods (breast cancer), salted food (stomach cancer), low fruits and vegetables intake (breast, oral, and lung cancer), red meat consumption (colon, breast, and pancreatic cancer) and low fiber diet (colon cancer) [23].

The risk of developing cancer is reduced by half in people consuming diets high in fruits and vegetables when compared with those consuming diets with less fruits and vegetables. The anticarcinogenic potential foods include vitamin C, vitamin A, vitamin E, selenium, allium plants, soybean, cruciferous vegetables, flax seeds and dietary fibers. The major vegetables include broccoli, cauliflower, radish, kale, Brussels sprouts, watercress, and cabbage that are used either fresh or cooked, but the lower incidences of many chronic diseases such as cancer and cardiovascular-related ailments are associated with consumption of vegetables rich dietary regimes [23–25].

High fiber diet prevents prostate cancer progression in the early stages based on Asian and Western cultures. Therefore, whole grain and bran products promise potential applications as anticancer ingredients in functional foods [26]. Vitamin B6 and riboflavin intakes from the diet were associated with a decreased risk of colorectal, lung, breast cancer and so on [27].

Several functional foods that have been obtained from various sources such as plants, animals, and microbial sources are reported to possess anticancer effects with varying mechanisms of action as antimetastatic, induction of apoptosis, antiproliferative, antiangiogenic, scavenging of free radicals, inhibition of matrix metalloproteinases, etc. Examples of functional foods that evinced anticancer potential include cereals, vegetables, beverages, dairy products, fish oil, beef, mushroom, and probiotics [28, 29].

Nutrition is related to about 30% of all the cancers cases. Cancer biologists have concerned in the application of natural products to improve the survival rate of cancer patients. Americans, Japanese, and Europeans are turning to the use of dietary vegetables, medicinal herbs, and their extracts or components to prevent or treat cancer [20].

Functional foods are foods and food components that supply health benefits beyond basic nutrition. These foods are similar in appearance to conventional foods; functional foods consumed as part of the normal diet. Functional food supplies the body with the needed amount of vitamins, fats, proteins, carbohydrates, required for its healthy survival [20].

There are several examples of functional foods showing anticancer activity that may be exhibited via varying modes of action, e.g., bitter melon inhibits the proliferation of cancer cells, fish oil exhibits anticancer activity via inhibition of metastasis, several species of mushrooms especially *Tricholoma matsutake* act by inducing apoptosis and has shown beneficial effects in the treatment of oral cancer. Some functional foods have shown anticancer efficacy via inhibition of pathological angiogenesis such as green tea, red grapes, etc. Several functional foods evinced anticancer activity via scavenging free radicals, for example, garlic, broccoli, green tea, soybean, tomato, carrot, cabbage, onion, cauliflower, red beets, cranberries, cocoa, blackberry, blueberry, red grapes, prunes, and citrus fruits. Thus, functional foods offer a potential for the prevention and treatment of cancer and therefore may act as an alternative for cancer therapy [28].

Wheat straw hold various bioactive compounds such as policosanols, phytosterols, phenolics, and triterpenoids, having enormous nutraceutical properties like anticancer, anti-inflammatory, anti-allergenic, antioxidant, anti-atherogenic, anti-microbial, anti-thrombotic, antiviral, cardioprotective, and vasodilatory effects [30]. β-glucans belong to a group of polysaccharides located in the cell wall of bacteria, fungi including mushrooms, as well as cereals such as barley and oats, and their anticancer effects were demonstrated mainly in in-vitro and in-vivo experimental systems [31].

Cereals (rice, barley, and wheat) have been reported to possess cancer preventive property. Its cancer preventive activity can be attributed to 'lunasin' a cancer preventive, anti-inflammatory, and cholesterol-reducing peptide found in cereals. Bran is the hard outer layer of cereal grains, which accounts for 10%, rich in a myriad of phytochemicals with anticancer as phenolics, flavonoids, glucans, and pigments. Twelve sphingolipids from wheat bran extract showed growth inhibition against human colon cancer cell lines in vitro. The antitumor effects of mushrooms have been reported for breast, colon, gastric, prostate, pancreatic, cervical, and ovarian cancer as well as endometrial cancer [32].

Rice (*Oryza sativa* L.) is one of the most important cereal for human nutrition [33] and contributes 21% to human's nutrient intake and energy requirements. Asia accounts for 92% of the world's total rice production (6.78 million tonnes). Germinated brown rice is not only richer in the basic nutritional and bioactive components, but also become a popular functional food, which exhibits many physiological effects, including anticancer, antihypertension, antidiabetes, etc. chronic diseases [34]. The pigmented rice contains a variety of flavones, tannin, phenolics, sterols, tocols, γ-oryzanols, amino acids, GABA, and essential oils, which has a lot of bioactivities including antitumor, antioxidant, antiatherosclerosis, hypoglycemic, and antiallergic activities [35]. The dietary rice bran may exert beneficial effects against several types of cancer, such as leukemia, breast, lung, liver, cervical, stomach, and colorectal cancer [36, 37]. Therefore, dietary rice bran and brown rice have the potential to have a significant impact on cancer prevention for the global population.

The anticancer activity can be mainly attributed to biopolymers. Higher consumption of fruit improves protection for maintaining human health against oxidative damage that may play a role in carcinogenesis and some chronic diseases, namely oxidative damage and lipid peroxidation. Citrus fruits have been reported to possess potential anti-cancer activity, which can be attributed to the presence of bioactive compounds like limonene and flavonoids such as quercetin, myricetin, rutin, tangeritin, naringin, and hesperidin. Pomegranate has been reported to be effective in the prevention of colon, prostate, and breast cancer because of hydroxycinnamic acids found in all parts of fruit [32].

Ellagic acid and resveratrol are found in grape skin, berries, and red wine that are known to have anti-inflammatory activity and inhibit the synthesis of prostaglandins stimulating tumor cell growth [32]. Probably the most intensively investigated class of physiologically-active components derived from animal products are the fatty acids, predominantly found in fatty fish such as salmon, tuna, mackerel, sardines, and herring.

There's a delicate balance of bioactive components in rice bran show anti-cancer activity, namely dietary rice bran may exert beneficial effects against breast, lung, liver, prostate, intestinal, and colorectal cancer. Diet with low fresh vegetables is a major factor that causes cancers, especially the deficiency of malnutrition. There are vitamins from vegetables with anticancer activity but they are easily destroyed after cooking [38].

Vitamin B6 and riboflavin intakes from the diet were associated with a decreased risk of colorectal, lung, and breast cancer [27]. Higher concentrations of plasma carotenoids and lutein may reduce the risk of urothelial cell carcinoma [39]. Selenium (Se) and calcium (Ca) deficiency were associated with cancer risk, but higher Se supplementation has been associated with lower cancer mortality [40], and a possible role for increasing dietary calcium intake in lung cancer prevention among female nonsmokers [41]. Iron (Fe) efficacy is a recommended key step in modern cancer patient management to minimize the impact on quality of life and performance status [42]. Zinc level regulation dysfunction has been identified in prostate cancer cells and may thus play an important role in the prostate cancer pathogenesis [43].

1.3 FUNCTIONAL FOOD FOR CARDIO-PROTECTIVE EFFECT

Numerous studies have confirmed that diets rich in fruits and vegetables are cardio-healthy and contain some bioactive substances that make food preventive against cardiovascular diseases; On the other hand, the prevalence of cardiovascular diseases is increasing around the world. It is important to address the risk factors related to this disease such as smoking, dyslipidemia, obesity, diabetes, hypertension, and ultimately hyperhomocysteinemia [44].

1.4 POLYPHENOLS

In nature, there are a wide variety of compounds that have a molecular structure characterized by the presence of one or more phenolic rings. These compounds can be called polyphenols (Figure 1.1). They originate mainly in plants, which synthesize them in great quantity, as a product of their secondary metabolism. Some are indispensable for physiological functions. Others participate in defensive functions in situations of stress and prevention of predators [45].

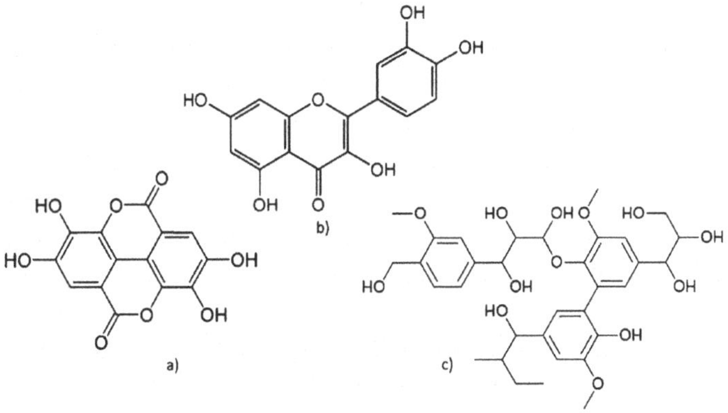

FIGURE 1.1 Example of main polyphenols, (a) hydrolyzable tannins, (b) flavonoid or condensed tannins, and (c) lignin.

1.5 CATECHINS AND PROANTHOCYANIDINS

The beneficial health effects of catechins and proanthocyanidins present in fruits and vegetables are related to their antioxidant properties (inhibition of LDL oxidation) as well as inhibition of platelet aggregation, modulation of endothelial function and antihypertensive properties. Experimental studies with isolated animals and tissues have shown that proanthocyanidins have a vaso-relaxing effect and this depends on nitric oxide-cyclic guanosine monophosphate (NO-cGMP) [44].

The epigallocatechin gallate characteristic of green tea composition reduces vascular inflammation by increasing nitric oxide (NO) synthesis, thereby blocking endothelial exocytosis [44]. On the other hand, it was demonstrated in an experimental study with rats; that at the dose of 50 mg/kg and 200 mg/kg showed lower levels of total cholesterol, LDL and CT/HDL and LDL/HDL significantly, however, triglyceride and HDL levels did not increase significantly [46].

Diet plays an important role in the maintenance of optimal cardiovascular health [47]. Consumption of fruits and vegetables is associated with lower concentrations of total and low-density lipoprotein cholesterol and with a decreased incidence of CVD [49]. This effect has been ascribed, in part to the antioxidants; Polyphenols, tocopherols, and carotenoids in plant-based foods. Plant foods rich in antioxidant molecules have received a growing interest because they delay the oxidative degradation of lipids [50, 51].

Another author in an experimental study concluded that the proanthocyanidins present in the serum of rats two hours after the administration of a grape seed extract, decreased the synthesis of triacylglycerols, cholesterol, and cholesterol-esters in cultures of human liver cells; on the other hand flavonoids which have a catechol group on ring B in their structure, a double bond between carbon 2 and carbon 3 of ring C and a ketone group on carbon 4 of ring C are the most effective inhibitors of the angiotensin converting enzyme; in turn the proanthocyanidins of grape seed extract had an antihypertensive effect in spontaneously hypertensive rats at a dose of 375 mg/kg, with a maximal effect at 6 hours after the ingestion of the extract, returning to the basal levels of tension arterial at 48 hours post administration [52].

Regular consumption of vegetables including tomatoes is associated with numerous health benefits, such as prevention of chronic-degenerative and cardiovascular diseases. Responsible for these properties in fruits, are the antioxidant compounds that comprise lycopene, flavonoids, phenols, and vitamins as C and E. The content and stability of these compounds depends on the cultivar or variety used, environmental conditions of cultivation, state fruit ripening and post-harvest treatments such as scalding, cutting, packaging, and refrigeration. Also, there are differences in the stability and bioavailability of the antioxidant compounds in the different tomato products, this information is important to diffuse the consumption of functional products [53].

Rice bran oil contains an array of bioactive phytochemicals of antioxidant activity such as oryzanols, phytosterols, tocopherols, tocotrienols, and policosanols [54, 55]. It was reported that rice bran oil possesses hypocholesterolemic properties primary attributable to its balanced fatty acid composition and high levels of the antioxidant phytochemicals [56] like orzynol, that has a greater effect on lowering plasma LDL-cholesterol levels and raising plasma HDL-cholesterol possibly through increasing the fecal excretion of cholesterol and its metabolites [56]. Wheat germ oil has been reported to contain phytosterols, unsaturated fatty acids and oryzanol which collectively possess lipid-lowering and antioxidant effects [51].

Carrot and tomato acetone extracts in the studied mixture are good sources of carotenoids, especially b-carotene and lycopene, respectively, which were reported previously to possess antioxidant and hypocholesterolemic effects [58, 59]. Lycopene and b-carotene augmented the activity of the macrophage LDL receptor [58]. In addition, tomato contains esculeoside A, a new glycoside, which was identified from tomato with contents 4 times higher than lycopene [51, 60].

The oral administration of esculeoside A to apodeficient mice significantly reduced the levels of serum cholesterol, triglycerides, LDL-cholesterol, and the areas of atherosclerotic lesions without any detectable side effects [60]. Epidemiological studies suggest that a diet rich in carotenoids is associated with a reduced risk of heart disease and cancer [61] Harari et al., [61] suggested that 9-cis b-carotene have the potential to inhibit atherogenesis in humans via its conversion to 9-cis retinoic acid, a ligand of the nuclear receptor Rexinoid x Receptor (RxR). Several studies indicated that RxR and its heterodimers have the potential to reduce atherosclerosis by affecting lipid metabolism [62], cell migration [63], apoptosis [64], and inflammation [51, 65].

1.6 ISOFLAVONES

Some studies mention that soy contains isoflavones, where the role of isoflavones is the reduction of serum cholesterol levels. Other clinical studies show that isoflavones reduce the susceptibility of reducing lipid oxidation and it has been observed that they may have a similar effect to digitalis in the relaxation of the coronary arteries through the mechanism that involves blocking the calcium channels [44].

Flavonoids may prevent the adhesion of monocytes in the inflammatory process of atherosclerosis, thereby having health benefits in cardiovascular disease. Flavonoid has been reported to offer protection against cardiovascular diseases, such as coronary heart disease [51, 66].

1.7 RESVERATROL AND CARDIOVASCULAR HEALTH

In wines is where resveratrol is found, especially in red wine, it is cardioprotective and acts on different levels. Some studies show the antiplatelet effect of resveratrol in vivo and in vitro; also has the ability of vasodilatation attributing the ability to stimulate the calcium and potassium channels and to improve nitric oxide in the endothelium [44].

1.8 MODULATOR OF LIPID METABOLISM

In vivo experiments in rats with spontaneous hypertension prone to myocardial infarction demonstrate that resveratrol reduces markers of oxidative stress in glycated serum and 8-hydroxyguanosine in urine albumin [44].

1.9 BIOACTIVE PEPTIDES

The main effect of peptides in the cardiovascular system describes that these are related to the antithrombotic and antihypertensive activity. Some of the peptides with greater antithrombotic activity are found in milk such as α-lactalbumin (α-LA), β-lactoglobulin (β-LG), caseins (CN), immunoglobulins (Ig), lactoferrin (LF), peptide-protein fractions, phosphoglycoproteins, and minor serum proteins (transferrin and albumin serum) [44].

1.10 PLANT STEROLS

There is a wealth of scientific evidence that sterols in plants have a significant hypocholesterolemic effect by reducing the concentration of both total cholesterol and LDL cholesterol (Low-Density Lipoproteins). The most studied effect of the sterol plants (Figure 1.2) is the inhibition of intestinal absorption of cholesterol, being able to move cholesterol from absorption micelles, reducing, and adding plant sterols, reducing the amount of cholesterol exported in the blood in the form of chylomicrons; the higher plants contain phytosterol which is actually a mixture of three component components: campesterol, sitosterol, and stigmasterol. Fungi and yeast contain ergosterol type sterols [44].

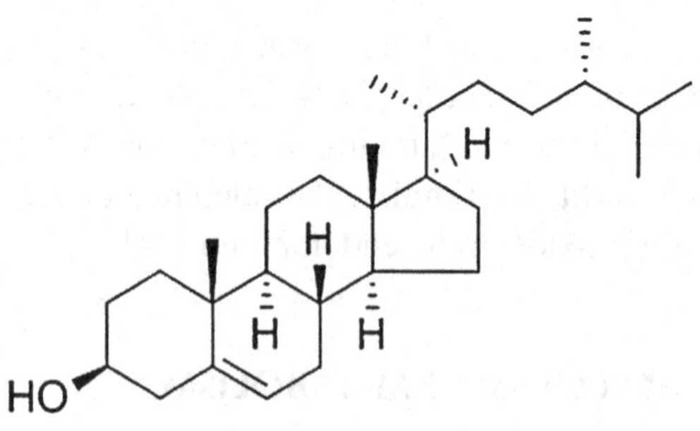

FIGURE 1.2 Campesterol as plant sterol example.

1.11 OMEGA–3 FATTY ACIDS

There is compelling evidence that different types of dietary fatty acid have divergent effects on CVD risk and that the type of fat is more important than the total amount of fat. The effect of a specific fatty acid depends strongly on the source of calories with which it is compared. Little or no cardiovascular benefits were seen when saturated fatty acids (SFA) is replaced by total carbohydrate, but a significant reduction in CVD risk is achieved when SFA is replaced by monounsaturated fatty acids (MUFA) and/or polyunsaturated fatty acids (PUFA). The effects of different food sources of SFA and different lengths of SFA warrant further study [67].

Although higher intake of trans fatty acids (TFA) from hydrogenated vegetable oils has adverse effects on lipid profile and CVD risk, the effects of trans isomers from ruminant fat remain controversial. Abundant evidence from randomized clinical trials (RCTs) and prospective cohort studies support the benefits of both *n*–6 and *n*–3 PUFAs, including both plant-sourced α-linoleic acid (ALA) and long-chain *n*–3 PUFAs from fish, in reducing CVD risk, but the benefits of fish oil supplements have not been clearly demonstrated [67].

Fatty acids, in addition to their known energy value, are part of the phospholipids of the membranes of the body cells, exerting a clear influence on the composition of the cell membrane and determining, to a greater or lesser degree, the structure and finality of the cell. This functionality includes various aspects such as fluidity and permeability, lipid peroxidation, gene influence, etc. [67].

The main sources of alpha-linolenic acid (ALA) are walnuts and especially the vegetable oils of flax, rapeseed, safflower, soybean, evening primrose and flax. As for eicosapentaenoic acid (EPA) and docosahexaenoic acid (DHA), the richest sources are fish oils. The content of omega-3 polyunsaturated fatty acids (PUFAs ω-3) varies according to the fish species, their location, the season of the year and the availability of phytoplankton [68].

The ω–3 PUFAs act on the cardiovascular system through a multitude of pathways exerting a beneficial effect on cardiovascular risk. They exert a stabilizing action of the cellular membrane producing an antiarrhythmic effect [68]. In addition, the ω–3 PUFAs inhibit platelet aggregation, particularly collagen-induced aggregation, and thromboxane A2 (TXA2) production, which discretely prolong bleeding time when administered at doses >3 g/day. They are also attributed overall favorable effects on the lipid profile (decrease in triglycerides and VLDL cholesterol, possible increase in HDL cholesterol) and hypotensive properties [68] (Figure 1.3).

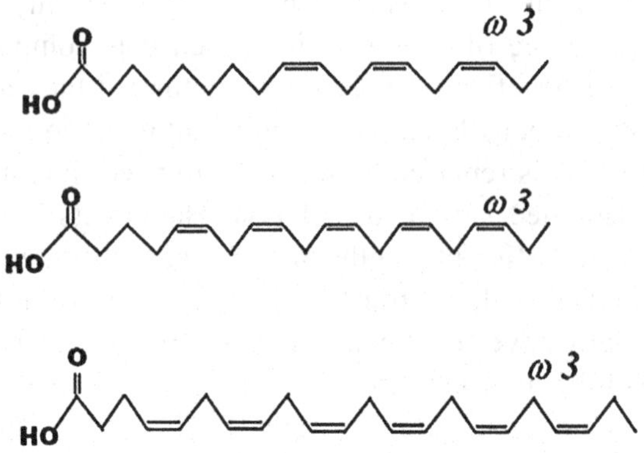

FIGURE 1.3 Examples of meaning ω–3 PUFAs: alpha-linolenic acid, eicosapentaenoic acid, and docosahexaenoic acid.

Several studies [68, 69] show that fish behaves as a more efficient vehicle in terms of bioavailability, in addition to providing proteins of high biological value and trace elements such as iodine and selenium [68].

The metabolic syndrome is of great concern worldwide because it increases enormously the risk of suffering type-2 diabetes mellitus (T2DM) and cardiovascular disease. Fortunately, epidemiological studies have shown that healthy diets rich in fruits, vegetables, grains, fish, and low-fat products have a protective role against the components of the metabolic syndrome, such as hypertension, hyperglycemia, hypertriglyceridemia, obesity, and low HDL levels. In this regard, dietary lipids are at the same time capable of developing these risk factors and to ameliorate them. SFA is overall deleterious, while MUFA and PUFA, as well as some of the minor components present in some dietary oils, are beneficial for these factors, despite some the different controversies raised [70].

In any case, modification of the fatty acid and phospholipid composition of plasma membrane leads to changes in the functionality of proteins involved in insulin activity, like the insulin receptor, GLUT-4, CD36/FAT and ABCA-1, which have important roles in the metabolism of glucose, fatty acids and cholesterol, and, in turn, in the key features of the metabolic syndrome. In consequence, modification of plasma membrane lipid composition with dietary lipids not only might protect against the development of insulin resistance, but also it can improve some of the altered markers of the metabolic syndrome. In fact, by modifying cell membrane composition, dietary oils are capable of reducing serum glucose and triglyceride levels, increasing those of HDL-cholesterol and lowering blood pressure [70].

As for differences by gender, men need to eat more fish than women to obtain the same level of PUFA ω–3 [68]. Table 1.1 describes the mechanism of action of some food functional ingredients.

1.12 FUNCTIONAL FOODS IN DIABETES AND OBESITY

As stated earlier, obesity is one of the principal NCD and a determinant risk factor for the development of other NCD as DM and cancer for example; DM is one of the most important causes of death and an NCD targeted as "priority" for the governments around the world [5]. DM is also intimately related to diet since one of its principal axes is the management of blood glucose concentration that depends on a big proportion of the quantity and type of sugar comprised in the diet. Due to it's various short and long-term complications, health researchers have been researched and developed a wide range of treatments including consuming of some foods claimed as functional. Since 2003, plants have been declared as a great source of dietary supplements that may represent a valuable help in control of DM and in the prevention of its complications [72]. The use of foods, principally plant-derived foods, as treatment for DM has its beginning in the traditional medicinal practices (also named "alternative medicine") generally raised from the religions practiced by ancestral cultures, an example is the Unani medicine from India where the pomegranate (*Punica granatum* L.) is used to control type 2 DM (T2D) [73]. During the past years, several studies have analyzed the relationship between diet content and T2D and found that consumption of fruits and vegetables blunted the risk of developing the illness [74]. These findings have generated a crescent interest of the scientific community on those foods, food ingredients (as raw material) or food components (as nutraceutical compounds) that have the possibility of exerts a beneficial activity on the people who present T2D and its complications.

TABLE 1.1 Meaning Compounds and Mechanism of Action of Some Functional Ingredients

Compound	Food	Beneficial effect	Mechanism of action	Reference
EPA/DHA	Fish oil	Diminution of heart attacks	Incorporation of EPA/DHA in membrane phospholipids of membrane de cardiomyocyte	[70]
NO-cGMP	Fruits/vegetables	Palette aggregation inhibition Endothelial function modulation Antihypertensive effect	Antioxidant properties (LDL oxidation inhibition)	[44]
Epigallocatechin gallate	Green tea	Decreases vascular inflammation	Nitric oxide (NO) synthase increases	[45]
Proanthocyanidins	Grape seeds extract	Hypercholesterolemia decreases	Nitric oxide (NO) synthase increases	[46]
Resveratrol	Red wine	Hearth protection Antiplatelet Vasodilatation Stimulation of calcium and potassium channels	Nitric oxide (NO) synthase increases	[47]
Linoleic acid	Fish/nuts	Decreases vascular inflammation	As part of phospholipids, it acts as a function-structure component	[48]

Many natural, unmodified foods have been tested to probe its potential antidiabetic and/or antiobesity functionality; these groups comprises vegetables, fruits or spices and have been found diverse results [75]. Most of these studies have as test material vegetables (understood as any eatable part of a plant) and fruits of dark colors or very strong tastes; some of these materials are blueberry [76, 77], blackberry [78, 79], grapes [80, 81], strawberries [82, 83], pomegranate [84, 85], cinnamon [86, 87], ginger [88], garlic [89, 90], among others. In the other hand, a wide different variety of bioactive compounds or food-obtained compounds/extracts (like essential oils or peel extracts) from foods have been used as potential functional foods ingredients or supplements instead unprocessed foods. Proteins like Casein [91], lipids as conjugated linoleic acid [92], organosulfur compounds like Z-ajoene [89], polyphenolic compounds like fisetin [93], resveratrol [94] or phenolic extract from different vegetal sources (even byproducts) [95–98] has been tested. These studies prove that virtually all fruits and vegetables are a good source of bioactive compounds with the ability to improve the metabolic stage of a diabetic subject.

1.13 *IN VITRO* STUDIES

All these studies have been carried out in all kind of experimental models. As an example epigallocatechin-3-gallate, the major phenolic compound found in green tea has been reported to present, besides its well known that antioxidant activity, a capacity to ameliorate the insulin resistance in a HepG2 cell model pretreated with high glucose in the culture medium to induce the impaired metabolic conditions [99]. Aloe vera, cinnamon, lychee, black cumin, avocado, grape, and ginger are foods that have been part of *in vitro* studies and showed the capacity to inhibit the activity of α-glucosidase and α-amylase [74].

1.14 ANIMAL STUDIES

During the recent years most research doing animal experimentation on the effect of any bioactive food/food component over DM, insulin resistance or metabolic syndrome prefers a model based on rats (*Ratus norvergicus*) with a variety of methods to induce the specific illness like alloxan or streptozotocin for DM [100, 101] or fructose administration for insulin resistance [102]; besides, some studies have been carried out in normal animals [74]. Other experiments used a genetically modified organism that expresses the desired illness [97]. Experiments on animals have thrown a very interesting result. Treatment of experimental diabetic animals with a variety of vegetable extracts (ethanolic, aqueous, hydroalcoholic, or organic solvent extracts), essential oils or plant parts (seed, bulbs) have thrown four principal effects: (i) Low of glucose concentration in plasma, serum or whole blood in fasting and/or postprandial conditions; (ii) Increments in the production, secretion and/or blood concentration of insulin; (iii) Improvement of lipid profile lowering the serum levels of total cholesterol and triglycerides and raising the concentration of High Density Lipoproteins (HDL); and (iv) Diminutions in glycosylated hemoglobin, that is a long-term reference of blood glucose levels [74]. Besides these benefits, protection of liver damage, recuperation of renal function [103, 104], regeneration of pancreatic β-cells [105, 106], and protection against weight loss proper of DM [107] are others beneficial effects exerted by treating diabetic animals with extract from onion, saffron, bitter gourd, black cumin and ginger [75]. Additionally, improving effects of the impaired protein metabolism of DM have been reported in animal studies too [108–110].

1.15 HUMAN STUDIES

Human trials have been carried out with diabetic, pre-diabetes, and metabolic syndrome patients under a wide variety of experimental designs. It is possible to found experiments so different from clinical trials with a small number of participants [111–113] to big randomized, double-blinded, and placebo-controlled multicenter protocols [114–116]. Nevertheless the size and experimental design used for each study, the overall results thrown by these works agree with the major effects found in the animal trails: (i) reduced levels of blood glucose (non fasted and fasting) as well as improved glucose tolerance; (ii) improved insulin metabolism, principally as an enhanced sensitivity to the hormone and decreased insulin resistance; and (iii) improved lipid profile, traduced as a reduced level of total cholesterol and triglycerides, lower concentration of low density lipoprotein and higher concentration of high density lipoproteins [74]. Additionally, reduced levels of fructosamine levels [113] and an improved function of β-cells [117] have been reported.

It is important to say that not all the studies report beneficial effects of the consumption of functional food or components *n* patients with DM. Although no toxic effect of these treatments has been reported, there are few studies where no difference is observed between the experimental group that consumes the functional food (*Momordica charantia*) and the control/placebo group [117, 118]. This lack of difference may be due to the fact that these studies administrated to patients capsules of the dried whole fruit of bitter melon and not an extract obtained from it. Virtually, all studies that report an effect of consumption of this functional fruit over metabolism have as treatment the administration of a variety of extracts (aqueous, ethanolic or other organic solvent extracts) [106, 111, 115, 119, 120]. Obviously, the administration of non-processed fruit (besides drying) have not the same presence of bioactive compounds that a treatment based on an extract obtained from the vegetal material that has been designed to be rich in those compounds. Whole non-treated or dried fruit may present problems of bioavailability or degraded content of the bioactive compounds that have the capability to exert the benefic effects.

1.16 ANTIOXIDANTS AND PROBIOTICS

It is evident that most reported functional foods are obtained from vegetables (or are the vegetables by themselves) but other important functional food components are probiotic. Probiotic was initially defined as compounds produced by microorganisms that promote growth in other species [121]; after that the term was defined as "a live microbial feed supplement which beneficially affects the host animal by improving its intestinal balance"; but nowadays, and according to the WHO, in the field of human health are known as live microorganisms, that when administrated in adequate amounts have benefic effect over the hosts health [122–124]. Healthy conditions of gut microbiota have been related in recent years to general health maintenance and specific diseases fight or prevention; number, type, and function of bacteria of microbiome vary among different parts of gastrointestinal tract and is viability depend on environmental factors as pH, oxygen, and composition of the ingested food [125]. Loss of microbiome homeostasis, known as dysbiosis, can be triggered by several factors as some drugs like antibiotics and others, chronic stress, gastrointestinal tract infections (disturbing typical bacteria species equilibrium) and, perhaps the most important, diet composition especially high-fat and high-protein diets. Dysbiosis can lead to a wide range of health problems including obesity and DM due to the tight relationship of the microbiome with the metabolism of energy and glucose.

In diet, two principal sources of probiotics are fermented beverages and dairy products; both can be considered as functional foods due to its beneficial microorganism content besides its content of other bioactive (as an antioxidant). Probiotics begin to be a point of interest since 1977, when Mann observed that consumption of big amounts of milk fermented with wild strains of *Lactobacillus* and *Bifidobacterium* (yogurt) provoked a diminution in serum cholesterol [126]. Since then, profilactic use of probiotics has been investigated as a dietary supplement or complementary treatment that can improve metabolic health much more far than gut microbiome. It has been proved that gut microbiota is intrinsically related to immune and inflammatory responses [127, 128]; this relationship directly involves the conditions of gut microbiome to DM and obesity through the inflammatory component of both diseases. Relation with DM is tightly bonded to the composition of microbiota; Larsen et al. demonstrated in 2010 that diabetic patients have a significantly lower proportion of *Firmicutes* and *Bifidobacteria* species in microbiota than non-diabetic subjects [129]. A more recent study reported that alloxan-induced diabetic rats feed with fermented milk containing a *Lactobacillus fermentum* strain (RS–2) were partially protected against the development of diabetic complications improving the glucose control and the antioxidative enzymatic profile compared with diabetic control group [130–133]. All these studies make an outlook on the state of the art of the functional foods as a treatment of the diabetes mellitus and the obesity.

1.17 CONCLUSIONS

There exist a series of food that due to his physicochemical components help to protect against the above-mentioned disease, inside the fruits and vegetables can be emphasized. In turn, also there are other groups of foods that consumed in excess can favor the appearance of cancer, particularly, the food with high contents of carcinogens, such as heavy metals, pesticides, and aromatic hydrocarbons present in smoked foods.

In general, it can be concluded that more than consuming a group of foods to prevent cancer, it is more important to follow a balanced diet, with a high consumption of fruits, vegetables, whole grains, legumes, nuts, and fish, which will cause Decrease the consumption of other foods not so beneficial. As a final conclusion, a balanced diet and an active lifestyle contribute beneficially to the nutritional status of the population that promotes the maintenance of good health, particularly in relation to the onset of cancer. In addition, it is necessary to emphasize the importance of the feeding not only as preventive of cancer but during the evolution of the same.

The consumption of polyunsaturated fatty acids ω–3 is related to a decrease in the risk of cardiovascular diseases, reducing the risk of death associated with this type of pathology. Fatty fish, rich in polyunsaturated fatty acids ω–3, besides being an excellent source of proteins and minerals, are presented as a reference food in cardio-healthy diets.

Resuming, many natural foods have the potential to be functional and many of them are capable to give a support for the treatment of metabolic diseases as DM, metabolic syndrome and obesity, mainly fruits and vegetables. These foods represent too a good source of bioactive compounds that can be used as a functional ingredient in the preparation of specially designed foods that can help to improve the life quality of diabetic and obese patients. By another hand, there exist alimentary products that present functionality over this illness like dairy and fermented foods (like cheese and wine, for example). The inclusion of these functional foods in the regular diet of diabetic and obese patients is a common practice but needs to be evaluated by health professionals to reach the possibility to gain maximum benefits from them.

KEYWORDS

- **cancer**
- **cardiovascular disease**
- **diabetes**
- **functional food**
- **non-communicable disease**

REFERENCES

1. Roberfroid, M. B., (2007). Inulin-type fructans: Functional food ingredients. *J. Nutr.*, *137*(11), 2493–2502.
2. Ashwell, M., (2002). *Concepts of Functional Foods*. Available at: http://ilsi.org/publication/concepts-of-functional-foods/. Accessed on September 3, 2017.
3. Pravst, I., (2012). Functional foods in Europe: A focus on health claims. In: Benjamin Valdez, (ed.), *Scientific, Health and Social Aspects of the Food Industry* pp. 165–208.
4. Della, V. E., Cacciatore, F., Farinaro, E., Salvatore, F., Marcantonio, R., Stranges, S. et al., (2017). The Mediterranean Diet in the prevention of degenerative chronic diseases. In: *Superfood and Functional Food* (pp. 115–132). INTECH.
5. WHO. World Health Organization, (2017). *Non-Communicable Diseases*. Available at: http://www.who.int/mediacentre/factsheets/fs355/en/.
6. Camps, J., García-Heredia, A., Hernández-Aguilera, A., & Joven, J., (2016). Paraoxonases, mitochondrial dysfunction, and non-communicable diseases. *Chem. Biol. Interact.*, *259*, 382–387.
7. WHO, (2013). *Global Action Plan for the Prevention and Control of Non-communicable diseases*, 1st ed., World Health Organization, Ed., Who library cataloging-in-publication data: Geneva.
8. GBD 2015 risk factor collaborators, (2016). Global, Regional, and National Comparative Risk Assessment of 79 Behavioural, Environmental and occupational, and metabolic risks or clusters of risks, 1990–2015: A systematic analysis for the global burden of disease study 2015. *Lancet*, *388* (10053), 1659–1724.
9. Keys, A., Menotti, A., Karvonen, M. J., Aravanis, C., Blackburn, H., Buzina, R. et al., (1986). The diet and 15-year death rate in the seven countries study. *Am. J. Epidemiol.*, *124*(6), 903–915.
10. Hoffman, R., & Gerber, M., (2012). The Mediterranean Diet : Health and Science, *Br. J. Nutr. 113*(2), 4–10.
11. Buil-Cosiales, P., Toledo, E., Salas-Salvadó, J., Zazpe, I., Farràs, M., Basterra-Gortari, F. J. et al., (2016). Association between dietary fiber intake and fruit, vegetable or whole-grain consumption and the risk of CVD: Results from the prevención con dieta mediterránea (predimed) trial. *Br. J. Nutr.*, *116*(3), 534–546.
12. Hilton, J., (2016). Growth patterns and emerging opportunities in nutraceutical and functional food categories: Market overview. In: *Developing New Functional Food and Nutraceutical Products* (pp. 1–28).
13. Nielsen, K. E., (2016). Health beneficial consumer products-status and trends. In: *Developing Food Products for Consumers with Specific Dietary Needs*, pp. 15–42.
14. Tapiero, H., Tew, K., Nguyen Ba, G., & Mathé, G., (2002). Polyphenols: Do they play a role in the prevention of human pathologies? *Biomed. Pharmacother.*, *56*(4), 200–207.
15. Sánchez-González, C., Ciudad, C. J., Noé, V., & Izquierdo-Pulido, M., (2017). Health benefits of walnut polyphenols: An exploration beyond their lipid profile. *CRC Crit Rev. Food Sci. Nutr.*, *57*(16), 3373–3383.

16. Arranz, S., Chiva-Blanch, G., Valderas-Martínez, P., Medina-Remón, A., Lamuela-Raventós, R. M., & Estruch, R., (2012). Wine, beer, alcohol and polyphenols on cardiovascular disease and cancer. *Nutrients*, 759–781.
17. Ayuso, M. I., Gonzalo-Gobernado, R., & Montaner, J., (2017). Neuroprotective diets for stroke. *Neurochemistry International*, 4–10.
18. Ogden, C. L., Carroll, M. D., Kit, B. K., & Flegal, K. M., (2012). Prevalence of obesity and trends in body mass index among US children and adolescents, 1999–2010. *JAMA*, *307*(5), 483–490.
19. Cerdá, C., Sánchez, C., Climent, B., Vázquez, A., Iradi, A., El Amrani, F. et al., (2014). Oxidative stress and DNA damage in obesity-related tumorigenesis. *Adv. Exp. Med. Biol.*, *824*, 5–17.
20. Aghajanpour, M., Nazer, M. R., Obeidavi, Z., Akbari, M., Ezati, P., & Kor, N. M., (2017). Functional foods and their role in cancer prevention and health promotion: A comprehensive review. *Am. J. Cancer Res.*, *7*(4), 740–769.
21. Turati, F., Bosetti, C., Polesel, J., Serraino, D., Montella, M., Libra, M. et al., (2017). Family history of cancer and the risk of bladder cancer. A case-control study from Italy. *Cancer Epidemiol.*, *48*, 29–35.
22. IOM/NAS. 1994. "Opportunities in the Nutrition and Food Sciences", Thomas, P. R. & Earl, R. (ed.). p. 109. Institute of Medicine/National Academy of Sciences, National Academy Press, Washington, D.C.
23. Bisen, P. S., (2016). Experimental and computational approaches in leveraging natural compounds for network-based anti-cancer medicine. *Cancer Med. Anticancer Drugs*, *1*(2), 1–3.
24. Brandt, K. E., Falls, K. E., Schoenfeld, J. D., Rodman, J. D., Gu, Z., Zhan, F. et al., (2018). Augmentation of intracellular iron using iron sucrose enhances the toxicity of pharmacological ascorbate in colon cancer cells, *Redox Biology, 14*, 82–87.
25. Mazewski, C., Liang, K., & Gonzalez de Mejia, E., (2018). Comparison of the effect of chemical composition of anthocyanin-rich plant extracts on colon cancer cell proliferation and their potential mechanism of action using in vitro, in silico, and biochemical assays, *Food Chem.*, *242*, 378–388.
26. Raina, K., Ravichandran, K., Rajamanickam, S., Huber, K. M., Serkova, N. J., Agarwal R., (2013). Inositol hexaphosphate inhibits tumor growth, vascularity, and metabolism in TRAMP mice: A multiparametric magnetic resonance study. *Cancer Prev. Res.*, *6*, 40–50.
27. Zschäbitz, S., Cheng, T. Y. D., Neuhouser, M. L., Zheng, Y., Ray, R. M., Miller, J. W. et al., (2013). B vitamin intakes and incidence of colorectal cancer: Results from the women's health initiative observational study cohort. *Am. J. Clin. Nutr.*, *97*(2), 332–343.
28. Gupta, P., (2016). Functional foods for cancer therapeutics. *Nat. Prod. Chem. Res. Gupta Nat. Prod. Chem. Res.*, *4*(2) 100-115.
29. Peng, Y., Meng, Q., Zhou, J., Chen, B., Xi, J., Long, P., Zhang, L. et al., (2018). Nanoemulsion delivery system of tea polyphenols enhanced the bioavailability of catechins in rats, *Food Chem.*, *242*, 527–532
30. Pasha, I., Saeed, F., Waqas, K., Anjum, F. M., & Arshad, M. U., (2013). Nutraceutical and functional scenario of wheat straw. *Critl. Rev. Food Sci. Nutr.*, *53*, 287–95.

31. Aleem, E., (2013). β-Glucans and their applications in cancer therapy: Focus on human studies. *Anticancer Agents Med. Chem.*, *13*(5), 709–719.
32. Patel, S., (2012). Cereal bran: The next super food with significant antioxidant and anticancer potential. *Mediterranean Journal of Nutrition and Metabolism*, pp. 91–104.
33. Huang, M. G., Narita, S., Numakura, K., Tsuruta, H., Saito, M., Inoue, T. et al., (2012). A high fat diet enhances proliferation of prostate cancer cells and activates MCP–1/CCR2 signaling. *Prostate*, *72*(16), 1779–1788.
34. Wu, F., Yang, N., Touré, A., Jin, Z., & Xu, X., (2013). Germinated brown rice and its role in human health. *Crit. Rev. Food Sci. Nutr.*, *53*(5), 451–463.
35. Deng, G. F., Xu, X. R., Zhang, Y., Li, D., Gan, R. Y., & Li, H. B., (2013). Phenolic compounds and bioactivities of pigmented rice. *Critl. Rev. Food Sci. Nutr.*, *53*(3), 296–306.
36. Chen, M. H., Choi, S. H., Kozukue, N., Kim, H. J., & Friedman, M., (2012). Growth-inhibitory effects of pigmented rice bran extracts and three red bran fractions against human cancer cells: Relationships with composition and antioxidative activities. *J. Agric. Food Chem.*, *60*(36), 9151–9161.
37. Henderson, A. J., Ollila, C.A., Kumar A., Borresen, E. C., Raina, K., Agarwal, R. et al., (2012). Chemopreventive properties of dietary rice bran: Current status and future prospects. *Adv. Nutr.*, *3*(5), 643–653.
38. Gaziano J. S. H., (2012). Multivitamins in the prevention of cancer in men: The physicians' health study ii randomized controlled trial. *JAMA J. Am. Med. Assoc.*, 1–10.
39. Ros, M. M., Bueno-de-Mesquita, H. B., Kampman, E., Aben, K. K., Buchner, F. L., Jansen, E. H. et al., (2012). Plasma carotenoids and vitamin C concentrations and risk of urothelial cell carcinoma in the European prospective investigation into cancer and nutrition. *Am. J. Clin. Nutr.*, *96*(4), 902–910.
40. Jaworska-Bieniek, K., Gupta, S., Durda, K., Muszynska, M., Sukiennicki, G., Jaworowska, E. et al., (2012). Microelements as risk factors for cancer of the lung and larynx. *Hered. Cancer Clin. Pract.*, *10*(4) 30–31.
41. Takata, Y., Shu, X. O., Yang, G., Li, H., Dai, Q., Gao, J. et al., (2012). Calcium intake and lung cancer risk among female nonsmokers: A report from the shanghai women's health study. *Cancer Epidemiol. Biomarkers Prev.*, *22*(1), 50–57.
42. Aapro, M., Österborg, A., Gascón, P., Ludwig, H., & Beguin, Y., (2012). Prevalence and management of cancer-related anemia, iron deficiency and the specific role of IV iron. *Ann. Oncol.*, 1954–1962.
43. Gumulec, J., Masarík, M., Krízková, S., Babula, P., Hrabec, R., Rovný, A. et al., (2011). Molecular mechanisms of zinc in prostate cancer. *Klin. Onkol.*, *24*(4), 249–255.
44. Mulero, J., Abellan, J., Zafrilla, P., Amores, D., & Hernandez Sanchez, P., (2015). Bioactive substances with preventive effect in cardiovascular diseases. *Nutr. Hosp.*, *32*(4), 1462–1467.
45. Pence, B. D., Bhattacharya, T. K., Park, P., Rytych, J. L., Allen, J. M., Sun, Y., McCusker, R. H., Kelley, K. W., Johnson, R. W., Rhodes, J. S., & Woods, J. A. (2017). Long-term supplementation with EGCG and beta-alanine decreases mortality but does not affect cognitive or muscle function in aged mice. *Exp Gerontol. 98*(1), 22–29.

46. Mohana, T., Navin, A. V., Jamuna, S., Sakeena M. S., & Devaraj S. N. (2015). Inhibition of differentiation of monocyte to macrophages in atherosclerosis by oligomeric proanthocyanidins *in-vivo* and *in-vitro* study. *Food Chem Toxicol. 82*(1), 96–105.
47. Iriti, M., Rossoni, M., & Faoro, F. (2006). Melatonin content in grape: myth or panacea? *J Sci Food Agric. 86*(10), 1432–1438.
48. Hennebelle, M., Zhang, Z., Metherel, A. H., Kitson, A.P, Otoki, Y., Richardson, C. E., Yang, J., Lee, K. S. S., Hammock, B. D., Zhang, L., Bazinet, R. P., & Taha, A. Y. (2017). Linoleic acid participates in the response to ischemic brain injury through oxidized metabolites that regulate neurotransmission. *Scientific Reports.* 4322(7), 2045–2322.
49. Quiñones, M., Miguel, M., & Aleixandre, A., (2012). Los polifenoles, compuestos de origen natural con efectos saludables sobre El sistema cardiovascular (Polyphenols, compounds of natural origin with healthy effects on the cardiovascular system). *Nutrición Hospitalaria,* 76–89.
50. Mezarina-Gómez, L. A., (2015). Hipolipidemic effect of the shell pulverizada of *Vitis vinifera* variety pinot noir (grape borgoña) in rats with induced hyperlipemia. *National University of San Marcos,* 24.
51. Dauchet, L., Amouyel, P., & Dallongeville, J., (2009). Fruits, vegetables, and coronary heart disease. *Nat. Rev. Cardiol., 6*(9), 599–608.
52. Mirmiran, P., Noori, N., Zavareh, M. B., & Azizi, F., (2009). Fruit and vegetable consumption and risk factors for cardiovascular disease. *Metabolism, 58*(4), 460–468.
53. Nicolle, C., Cardinault, N., Gueux, E., Jaffrelo, L., Rock, E., Mazur, A. et al., (2004). Health effect of vegetable-based diet: Lettuce consumption improves cholesterol metabolism and antioxidant status in the rat. *Clin. Nutr., 23*(4), 605–614.
54. Mohamed, D. A., Hamed, T. E., & Al-Okbi, S. Y., (2010). Reduction in hypercholesterolemia and risk of cardiovascular diseases by mixtures of plant food extracts: A study on plasma lipid profile, Oxidative Stress, and Testosterone in Rats. *Grasas y Aceites., 61*(4), 378–389.
55. Guerrero-Orjuela, L. S., (2013). Evaluation of the correcting effect of lipid synthesis and hypertension of extracts rich in polyphenols. Universitat Rovira I Virgili., 237.
56. Luna-Guevara, M. L., & Delgado, Alvarado, A., (2014). Importance, contribution and stability of antioxidants in fruits and products of tomato: Scientific information system. *Network of Scientific Journals of Latin America and the Caribbean, Spain and Portugal,* p. 63.
57. Khatoon, S., & Gopalakrishna, A. G., (2004). Fat-soluble nutraceuticals and fatty acid composition of selected Indian rice varieties. *J. Am. Oil Chem. Soc., 81*(10), 939–943.
58. Ardiansyah, S. H., Koseki, T., Ohinata, K., Hashizume, K., & Komai, M., (2006). Rice bran fractions improve blood pressure, lipid profile, and glucose metabolism in stroke-prone spontaneously hypertensive rats. *J. Agric. Food Chem., 54*(5), 1914–1920.
59. Devi, R. R., Jayalekshmy, A., & Arumughan, C., (2008). Antioxidant efficacy of phytochemical extracts from defatted rice bran in *in-vitro* model emulsions. *Int. J. Food Sci. Technol., 43*(5), 878–885.
60. Wilson, T. A., Nicolosi, R. J., Woolfrey, B., & Kritchevsky, D., (2007). Rice bran oil and oryzanol reduce plasma lipid and lipoprotein cholesterol concentrations and

aortic cholesterol ester accumulation to a greater extent than ferulic acid in hypercholesterolemic Hamsters. *J. Nutr. Biochem.*, *18*(2), 105–112.
61. Fuhrman, B., Elis, A., & Aviram, M., (1997). Hypocholesterolemic effect of lycopene and beta-carotene is related to suppression of cholesterol synthesis and augmentation of LDL receptor activity in macrophages. *Biochem. Biophys. Res. Commun.*, *233*(3), 658–662.
62. Elson, C., Peffley, D., Hentosh, P., & Mo, H., (1999). Isoprenoid-mediated inhibition of mevalonate synthesis: Potential application to cancer. *Proc. Soc. Exp. Biol. Med.*, *221*(4), 294–311.
63. Fujiwara, Y., Takaki, A., Uehara, Y., Ikeda, T., Okawa, M., Yamauchi, K. et al., (2004). Tomato steroidal alkaloid glycosides, Esculeosides A and B, from Ripe Fruits. *Tetrahedron*, *60*(22), 4915–4920.
64. Harari, A., Harats, D., Marko, D., Cohen, H., Barshack, I., Kamari, Y. et al., (2008). A 9-cis beta-carotene-enriched diet inhibits atherogenesis and fatty liver formation in LDL receptor knockout mice. *J. Nutr.*, *138*(10), 1923–1930.
65. Chinetti, G., Lestavel, S., Bocher, V., Remaley, A. T., Neve, B., Torra, I. P. et al., (2001). PPAR-Alpha and PPAR-gamma activators induce cholesterol removal from human macrophage foam cells through stimulation of the ABCA1 pathway. *Nat. Med.*, *7*(1), 53–58.
66. Day, R. M., Lee, Y. H., Park, A. M., & Suzuki, Y. J., (2006). Retinoic acid inhibits airway smooth muscle cell migration. *Am. J. Respir. Cell Mol. Biol.*, *34*(6), 695–703.
67. Ji, J. D., Cheon, H., Jun, J. B., Choi, S. J., Kim, Y. R., Lee, Y. H. et al., (2001). Effects of peroxisome proliferator-activated receptor-gamma (PPAR-Gamma) on the expression of inflammatory cytokines and apoptosis induction in rheumatoid synovial fibroblasts and monocytes. *J. Autoimmun.*, *17*(3), 215–221.
68. Marx, N., Sukhova, G. K., Collins, T., Libby, P., & Plutzky, J., (1999). PPAR-alpha activators inhibit cytokine-induced vascular cell adhesion molecule–1 expression in human endothelial cells. *Circulation*, *99*(24), 3125–3131.
69. Goldberg, D. M., Hahn, S. E., & Parkes, J. G., (1995). Beyond alcohol: Beverage consumption and cardiovascular mortality. *Clin. Chim. Acta.*, *237*(1–2), 155–187.
70. Wang, D. D., & Hu, F. B., (2017). Dietary fat and risk of cardiovascular disease: Recent controversies and advances. *Ann. Rev. of Nut.*, *37*, 423–446.
71. Piñeiro-Corrales, G., Lago Rivero, N., & Culebras-Fernández, J. M., (2013). Role of omega-3 fatty acids in cardiovascular disease prevention. *Nutr. Hosp.*, *28*(1), 1–5.
72. Harris, W. S., Pottala, J. V., Sands, S. A., & Jones, P. G., (2007). Comparison of the effects of fish and fish-oil capsules on the N 3 fatty acid content of blood cells and plasma phospholipids. *Am. J. Clin. Nutr.*, *86*(6), 1621–1625.
73. Perona, J. S., (2017). Membrane lipid alterations in the metabolic syndrome and the role of dietary oils. *J. Biochimica et Biophysica Acta (BBA) – Biomembranes. Book Review*, *1859*(9), Part B, 1690–1703.
74. Metcalf, R. G., James, M. J., Gibson, R. A., Edwards, J. R. M., Stubberfield, J., Stuklis, R. et al., (2007). Effects of fish-oil supplementation on myocardial fatty acids in humans. *Am. J. Clin. Nutr.*, *85*(5), 1222–1228.

75. Gallagher, A. M., Flatt, P. R., Duffy, G., & Abdel-Wahab, Y. H. A., (2003). The effects of traditional antidiabetic plants on *in vitro* glucose diffusion. *Nutrition Research*, pp. 413–424.
76. Ben, N. C., Ayed, N., & Metche, M., (1996). Quantitative determination of the polyphenolic content of pomegranate peel. *Zeitschrift für Leb. und Forsch.*, *203*(4), 374–378.
77. Beidokhti, M. N., & Jäger, A. K., (2017). Review of antidiabetic fruits, vegetables, beverages, oils and spices commonly consumed in the diet. *J. Ethnopharmacol.*, *201*, 26–41.
78. Prior, R. L., Wilkes, S., R. Rogers, T., Khanal, R. C., Wu, X., & Howard, L. R., (2010). Purified blueberry anthocyanins and blueberry juice alter development of obesity in mice fed an obesogenic high-fat diet. *J. Agric. Food Chem.*, *58*(7), 3970–3976.
79. Seymour, E. M., Tanone, I. I., Urcuyo-Llanes, D. E., Lewis, S. K., Kirakosyan, A., Kondoleon, M. G. et al., (2011). Blueberry intake alters skeletal muscle and adipose tissue peroxisome proliferator-Activated receptor activity and reduces insulin resistance in obese rats. *J. Med. Food*, *14*(12), 1511–1518.
80. Mykkanen, O. T., Huotari, A., Herzig, K. H., Dunlop, T. W., Mykknen, H., & Kirjavainen, P. V., (2014). Wild blueberries (Vaccinium myrtillus) alleviate inflammation and hypertension associated with developing obesity in mice fed with a high-fat diet. *PLoS One*, *9*(12) 1-21.
81. Kaume, L., Howard, L. R., & Devareddy, L., (2012). The blackberry fruit: A review on its composition and chemistry, metabolism and bioavailability, and health benefits. *J. Agric. Food Chem.*, *60*(23), 5716–5727.
82. Zanotti, I., Dall'Asta, M., Mena, P., Mele, L., Bruni, R., Ray, S. et al., (2015). Athero protective effects of (Poly)phenols: A focus on cell cholesterol metabolism. *Food Funct.*, *6*(1), 13–31.
83. Oueslati, N., Charradi, K., Bedhiafi, T., Limam, F., & Aouani, E., (2016). Protective effect of grape seed and skin extract against diabetes-induced oxidative stress and renal dysfunction in virgin and pregnant rat. *Biomed. Pharmacother.*, *83*, 584–592.
84. Almomen, S. M. K., Guan, Q., Liang, P., Yang, K., Sidiqi, A. M., Levin, A. et al., (2017). Daily intake of grape powder prevents the progression of kidney disease in obese type 2 diabetic ZSF1 rats. *Nutrients*, *9*(4), 1–16.
85. McDougall, G. J., Kulkarni, N. N., & Stewart, D., (2008). Current developments on the inhibitory effects of berry polyphenols on digestive enzymes. *BioFactors*, *34*(1), 73–80.
86. Basu, A., Betts, N. M., Nguyen, A., Newman, E. D., Fu, D., & Lyons, T. J., (2014). Freeze-dried strawberries lower serum cholesterol and lipid peroxidation in adults with abdominal adiposity and elevated serum lipids. *J. Nutr.*, *144*(6), 830–837.
87. Ok, E., Do, G. M., Lim, Y., Park, J. E., Park, Y. J., & Kwon, O., (2013). Pomegranate vinegar attenuates adiposity in obese rats through coordinated control of AMPK signaling in the liver and adipose tissue. *Lipids Health Dis.*, *12*(1), 163.
88. Park, J. E., Kim, J. Y., Kim, J., Kim, Y. J., Kim, M. J., Kwon, S. W., & Kwon, O., (2014). Pomegranate vinegar beverage reduces visceral fat accumulation in association with AMPK activation in overweight women. A double-blind, randomized, and placebo-controlled trial. *J. Funct. Foods*, *8*, 274–281.

89. Fabian, E., Töscher, S., Elmadfa, I., & Pieber, T. R., (2011). Use of complementary and alternative medicine supplements in patients with diabetes mellitus. *Ann. Nutr. Metab.*, *58*(2), 101–108.
90. Okutan, L., Kongstad, K. T., Jäger, A. K., & Staerk, D., (2014). High-resolution α-amylase assay combined with high-performance liquid chromatography–solid-phase extraction–nuclear magnetic resonance spectroscopy for expedited identification of α-amylase inhibitors: Proof of concept and α-amylase inhibitor in cinnamon. *J. Agric. Food Chem.*, *62*(47), 11465–11471.
91. Li, Y., Tran, V. H., Duke, C. C., & Roufogalis, B. D., (2012). Preventive and protective properties of *Zingiber officinale* (Ginger) in diabetes mellitus, diabetic complications, and associated lipid and other metabolic disorders: A brief review. *Evidence-based Complement. Altern. Med.*, 1–10.
92. Lee, D. Y., Li, H., Lim, H. J., Lee, H. J., Jeon, R. et al., (2012). Anti-Inflammatory activity of sulfur-containing compounds from garlic. *J. Med. Food*, *15*(11), 992–999.
93. Moradabadi, L., Montasser Kouhsari, S., & Fehresti Sani, M., (2013). Hypoglycemic effects of three medicinal plants in experimental diabetes: Inhibition of rat intestinal α-glucosidase and enhanced pancreatic insulin and cardiac glut-4 mRNAs expression. *Iran. J. Pharm. Res.*, *12*(3), 385–397.
94. Niamh, P., Nongonierma, A. B., Keane, D., Kelly, S., Celkova, L., Lyons, C. L. et al., (2016). A casein hydrolysate protects mice against high fat diet induced hyperglycemia by attenuating NLRP3 inflammasome-mediated inflammation and improving insulin signaling, 2421–2432.
95. Fuke, G., & Nornberg, J. L., (2017). Systematic evaluation on the effectiveness of conjugated linoleic acid in human health. *Crit. Rev. Food Sci. Nutr.*, *57*(1), 1–7.
96. Prasath, G. S., & Subramanian, S. P., (2011). Modulatory effects of fisetin, a bioflavonoid, on hyperglycemia by attenuating the key enzymes of carbohydrate metabolism in hepatic and renal tissues in streptozotocin-induced diabetic rats. *Eur. J. Pharmacol.*, *668*(3), 492–496.
97. Bo, S., Ponzo, V., Evangelista, A., Ciccone, G., Goitre, I., Saba, F., Procopio, M. et al., (2017). Effects of 6 months of resveratrol versus placebo on pentraxin 3 in patients with type 2 diabetes mellitus: A double-blind randomized controlled trial. *Acta Diabetol.*, 499–507.
98. Lin, D., Xiao, M., Zhao, J., Li, Z., Xing, B., Li, X. et al., (2016). An overview of plant phenolic compounds and their importance in human nutrition and management of type 2 diabetes. *Molecules*, *21*(10) 1-19.
99. Béjaoui, A., Ben Salem, I., Rokbeni, N., M'rabet, Y., Boussaid, M., & Boulila, A., (2017). Bioactive compounds from *hypericum humifusum* and *hypericum perfoliatum* : Inhibition potential of polyphenols with acetylcholinesterase and key enzymes linked to type–2 diabetes. *Pharm. Biol.*, *55*(1), 906–911.
100. Chen, S. J., Aikawa, C., Yoshida, R., Kawaguchi, T., & Matsui, T., (2017). Anti-prediabetic effect of rose hip (*Rosa canina*) extract in spontaneously diabetic torii rats. *J. Sci. Food Agric.*, *97*(12), 3923–3928.
101. Lao, F., & Sigurdson, G. T., (2017). Health benefits of purple corn (*Zea Mays* L.) phenolic compounds, *16*, 234–246.

102. Ma, S. B., Zhang, R., Miao, S., Gao, B., Lu, Y., Hui, S. et al., (2017). Epigallocatechin–3-gallate ameliorates insulin resistance in hepatocytes. *Mol. Med. Rep.*, *15*(6) 3803-3809.
103. Vanitha, P., Uma, C., Suganya, N., Bhakkiyalakshmi, E., Suriyanarayanan, S., Gunasekaran, P. et al., (2014). Modulatory effects of morin on hyperglycemia by attenuating the hepatic key enzymes of carbohydrate metabolism and β-cell function in streptozotocin-induced diabetic rats. *Environ. Toxicol. Pharmacol.*, *37*(1), 326–335.
104. Manjunath, K., Bhanu Prakash, G., Subash, K. R., Tadvi, N. A., Manikanta, M., & Umamaheswara Rao, K., (2016). Effect of aloe vera leaf extract on blood glucose levels in alloxan induced diabetic rats. *Natl. J. Physiol. Pharm. Pharmacol.*, *6*(5) 471–474.
105. Jalal, R., Bagheri, S. M., Moghimi, A., & Rasuli, M. B., (2007). Hypoglycemic effect of aqueous shallot and garlic extracts in rats with fructose-induced insulin resistance. *J. Clin. Biochem. Nutr.*, *41*(3), 218–223.
106. El-Demerdash, F. M., Yousef, M. I., & El-Naga, N. I. A., (2005). Biochemical study on the hypoglycemic effects of onion and garlic in alloxan-induced diabetic rats. *Food Chem. Toxicol.*, *43*(1), 57–63.
107. Elgazar, A. F., Rezq, A. A., & Bukhari, H. M., (2013). Anti-hyperglycemic effect of saffron extract in alloxan-induced diabetic rats. *Eur. J. Biol. Sci.*, *5*(1), 14–22.
108. Kanter, M., Meral, I., Yener, Z., Ozbek, H., & Demir, H., (2003). Partial regeneration/proliferation of the β-cells in the islets of langerhans by nigella Sativa L. in streptozotocin-induced diabetic rats. *Tohoku J. Exp. Med.*, *201*(4), 213–219.
109. Singh, N., & Gupta, M., (2007). Regeneration of β cells in islets of langerhans of pancreas of alloxan diabetic rats by acetone extract of *Momordica charantia* (Linn.) (Bitter Gourd) fruits. *Indian J. Exp. Biol.*, *45*(12).
110. Thompson, M., Al-Amin, Z. M., Al-Qattan, K. K., & Ali, M., (2007). Hypoglycemic effects of ginger in mildly and severely diabetic rats. *FASEB J.*, *21*(5), 50.
111. Al-Amin, Z. M., Thomson, M., Al-Qattan, K. K., Peltonen-Shalaby, R., & Ali, M., (2006). Anti-diabetic and hypolipidaemic properties of ginger (*Zingiber officinale*) in streptozotocin-induced diabetic rats. *Br. J. Nutr.*, *96*(4), 660–666.
112. Thomson, M., Al-Amin, Z. M., Al-Qattan, K. K., Shaban, L. H., & Ali, M., (2007). Anti-diabetic and hypolipidaemic properties of garlic (Allium Sativum) in streptozotocin-induced diabetic rats. *Int. J. Diabetes Metab.*, *15*, 108–115.
113. Saravanan, G., Ponmurugan, P., Kumar, G. P. S., & Rajarajan, T., (2009). Antidiabetic properties of S-allyl cysteine, a garlic component on streptozotocin-induced diabetes in rats. *J. Appl. Biomed.*, *7*(3), 151–159.
114. Leatherdale, B. A., Panesar, R. K., Singh, G., Atkins, T. W., Bailey, C. J., & Bignell, A. H., (1981). Improvement in glucose tolerance due to *Momordica charantia* (Karela). *Clin. Res.*, *282*(6279), 1823–1824.
115. Lerman-Garber, I., Ichazo-Cerro, S., Zamora-González, J., Cardoso-Saldaña, G., & Posadas-Romero, C., (1994). Effect of a high-monounsaturated fat diet enriched with avocado in NIDDM patients. *Diabetes Care*, *17*(4), 311–315.
116. Rosenblat, M., Hayek, T., Aviram, M., Rorke, E. A., Hughes, D. A., & Tarkum, I., (2006). Anti-oxidative effects of pomegranate juice (PJ) consumption by diabetic patients on serum and on macrophages. *Atherosclerosis*, *187*(2), 363–371.

117. Crawford, P., (2009). Effectiveness of cinnamon for lowering hemoglobin A1C in patients with type 2 diabetes: A randomized, controlled trial. *J. Am. Board Fam. Med.*, *22*(5), 507–512.
118. Fuangchan, A., Sonthisombat, P., Seubnukarn, T., Chanouan, R., Chotchaisuwat, P., Sirigulsatien, V. et al., (2011). Hypoglycemic effect of bitter melon compared with metformin in newly diagnosed type 2 diabetes patients. *J. Ethnopharmacol.*, *134*(2), 422–428.
119. Chuengsamarn, S., Rattanamongkolgul, S., Luechapudiporn, R., Phisalaphong, C., & Jirawatnotai, S., (2012). Curcumin extract for prevention of type 2 diabetes. *Diabetes Care*, *35*(11), 2112–2127.
120. John, J., Subhash, H. S., & Cherian, A. M., (2003). Evaluation of the efficacy of bitter gourd (*Momordica charantia*) as an oral hypoglycemic agent – a randomized controlled clinical trial. *Indian J. Physiol. Pharmacol.*, *47*(3), 363–365.
121. Dans, A. M. L., Villarruz, M. V. C., Jimeno, C. A., Javelosa, M. A. U., Chua, J., Bautista, R. et al., (2007). The effect of momordica charantia capsule preparation on glycemic control in type 2 diabetes mellitus needs further studies. *J. Clin. Epidemiol.*, *60*(6), 554–559.
122. Hafizur, R. M., Kabir, N., & Chishti, S., (2011). Modulation of pancreatic β-cells in neonatally streptozotocin-induced type 2 diabetic rats by the ethanolic extract of *Momordica charantia* fruit pulp. *Nat. Prod. Res.*, *25*(4), 353–367.
123. Chang, C. I., Chou, C. H., Liao, M. H., Chen, T. M., Cheng, C. H., Anggriani, R. et al., (2015). Bitter melon triterpenes work as insulin sensitizers and insulin substitutes in insulin-resistant cells. *J. Funct. Foods*, *13*(1), 214–224.
124. Walters, A. H., (1984). Probiotics in modern meat production. *Perspect. Public Health*, *104*(6), 220–224.
125. FAO; WHO, (2002). *Guidelines for the Evaluation of Probiotics in Food*; London, Ontario, Canada.
126. Chua, K. J., Kwok, W. C., Aggarwal, N., Sun, T., & Chang, M. W., (2017). Designer probiotics for the prevention and treatment of human diseases. *Curr. Opin. Chem. Biol.*, *40*, 8–16.
127. Esmaeili, S. A., Mahmoudi, M., Momtazi, A. A., Sahebkar, A., Doulabi, H., & Rastin, M., (2017). Tolerogenic probiotics: Potential immunoregulators in systemic lupus erythematosus. *J. Cell. Physiol.*, *232*(8), 1994–2007.
128. Pokrzywnicka, P., & Gumprecht, J., (2015). Intestinal microbiota and its relationship with diabetes and obesity. *Pediatr. Res.*, *78*(3), 232–238.
129. Enck, P., Zimmermann, K., Rusch, K., Schwiertz, A., Klosterhalfen, S., & Frick, J. S., (2009). The effects of maturation on the colonic microflora in infancy and childhood. *Gastroenterol. Res. Pract.*, 752401.
130. Kirjavainen, P. V., & Gibson, G. R., (1999). Healthy gut microflora and allergy: Factors influencing development of the microbiota. *Ann. Med.*, *31*(4), 288–292.
131. Gill, H. S., Rutherfurd, K. J., Prasad, J., & Gopal, P. K., (2000). Enhancement of natural and acquired immunity by Lactobacillus rhamnosus (HN001), Lactobacillus acidophilus (HN017) and Bifidobacterium lactis (HN019). *Br. J. Nutr.*, *83*, 167–176.

132. Larsen, N., Vogensen, F. K., Van Den Berg, F. W. J., Sandris Nielsen, D., Andreasen, A. S., Pedersen, B. K. et al., (2010). Gut microbiota in human adults with type 2 diabetes differs from non-diabetic adults. *PLoS One*, *5*(2), 1–190.
133. Kumar, N., Kumar Tomar, S., Thakur, K., & Kumar Singh, A., (2017). The ameliorative effects of probiotic lactobacillus fermentum strain RS–2 on alloxan induced diabetic rats. *J. Funct. Foods.*, *28*, 275–284.

CHAPTER 2

SYNERGISM IN FOOD TRENDS: THE KEY TO A HEALTHY LIFE

DULCE A. FLORES-MALTOS and JOSÉ A. TEIXEIRA

Institute for Biotechnology and Bioengineering, Centre of Biological Engineering, University of Minho, 4710-057 Braga, Portugal, E-mail: abril.maltos@ceb.uminho.pt

ABSTRACT

In this chapter, some of the newest trends in the incorporation of prebiotics and probiotics in food processing are reviewed, as well as the synergism between them and action for the benefit of the consumer. This revision intends to provide an updated overview on the development of functional product and the importance of sensory analytical techniques in the development of them. And also review how the incorporate of probiotics and prebiotics improves the quality of food and how it acts in the health and well-being of the consumer, including probiotics and prebiotics in the market, functional food manufacturing, preservation, and control.

2.1 INTRODUCTION

The primary role of diet is to provide enough nutrients to meet metabolic requirements, while giving the consumer a feeling of satisfaction, state of physical and mental well-being [1–4]. Foods contain various dietary components with an array of health benefits that offers an excellent opportunity to improve public health and well-being. Moreover, in recent years, consumers awareness towards the relationship between food and health has led to an explosion of interest in "healthy foods," this phenomenon could be partly attributed to the increasing cost of healthcare, the steady increase in life expectancy, and the desire of older people for an improved quality of their later years [4].

Nowadays, healthy foods mean "functional food," and we generally label a food as functional if it exerts beneficial effects or more specific body functions, in addition to the traditional nutritional effects [4].

Functional food is a part of the human diet and is demonstrated to provide health benefits and decrease the risk of chronic diseases beyond those provided by adequate nutrition [4].

Recent knowledge, however, supports the hypothesis that, beyond meeting nutrition needs, diet may modulate various physiological functions and may play detrimental or beneficial roles in some diseases [3].

There is a threshold between of a food technologies and nutrition sciences indeed, at least in the Western world, concepts are expanding on the survival, hunger satisfaction, and preventing adverse effects to an emphasis on the use of foods how tool to promote a well-being state, improving health, and reducing the risk of diseases [3].

A functional food must meet the following characteristics: (i) usual foods with naturally occurring bioactive substances (e.g., dietary fiber), (ii) foods supplemented with bioactive substances (e.g., probiotics, antioxidants), and (iii) derived food ingredients introduced to conventional foods (e.g., prebiotics) [5].

Nowadays established that the colonic microflora has a profound influence on health. Consequently, there is currently a great deal of interest in the use of prebiotic oligosaccharides as functional food ingredients to manipulate the composition and metabolism of colonic microflora in order to improve health [6].

2.2 PROBIOTIC

There are many proposed definitions for probiotics depending on their mechanisms of action and their effects on human health [7]. The concept of probiotic has been introduced maybe by the Russian Nobel laureate Elie Metchnikoff in 1907 ("The Prolongation of Life: Optimistic Studies") where he proposed the idea that ingesting microbes could have beneficial effects for human beings, especially to treat digestive diseases [8]. Nowadays, probiotic is defined as a live microbial food ingredient that when administered in adequate amounts confer a health benefit on the host [8–10].

Bacteria the genera *Bifidobacterium* and *Lactobacillus* are most used as probiotic supplements in processed food. These microorganisms exert only beneficial properties in the human and the animal health, such as inhibition of growth of exogenous and/or harmful bacteria, stimulation of immune system, anti-tumor properties, cholesterol reduction, and aid in digestion and/or absorption of food ingredients/minerals and synthesis of vitamins [11, 12]. The microorganisms currently used for their probiotic effect are shown in Table 2.1 and are mostly in the genera *Lactobacillus* and *Bifidobacterium*. However, other bacteria and some yeasts may have probiotic properties as well [13].

Immunological stimulation and resistance to pathogenic micro-organisms can be achieved through probiotic therapy. However, it is important that probiotics reach the large intestine intact and still viable. Upon ingestion, these probiotics are confronted by several physical and chemical barriers such as gastric and bile acids. At the end of the gastric route, only a small portion of probiotics are able to reach and settle in the colon [14].

Recent research focuses on the use of prebiotics as a vehicle because they are not viable in the first parts of the gastric system, but rather their function is to be a substrate for growth of the beneficial intestinal flora [15].

2.3 PREBIOTIC

Prebiotics differ from probiotics in that they do not contain live microbes, but stimulate their growth in the intestine [16]. The term prebiotic was first introduced by Gibson & Roberfroid in 1995 and is defined as a non-digestible, but fermentable, food ingredient that beneficially affects the host by selectively stimulating the growth and/or activity of one species or a limited number of species bacteria in the colon and thus improves host health [8, 9, 17, 18].

Prebiotics are more recent discovery, utilized to promote the survival of probiotics. Prebiotics are non-digestible carbohydrates that are not absorbed in the intestine. Prebiotics are generally oligosaccharides, whose degree of polymerization ranges between 2 and 20 monomers, they travel

TABLE 2.1 Microorganisms Used as Probiotics

Genus	Microorganisms
Lactobacilli[a]	*L. acidophilus*-group
	L. acidophilus (LA–5)
	L. crispatus (*L. acidophilus* "Gilliland")
	L. johnsonii (LA1)
	L .gasseri (PA 16/8)
	L. casei-group
	L(para)*casei* (*L.casei*) "shirota"
	L. casei "defensis"
	L. rhamnosus (LGG)
	L. reuteri
	L. plantarum (299 and 299v)
Bifidobacteria	*B. longum* (BB536)
	B. longum (SP 07/3)
	B. bifidum (MF 20/5)
	B. infantis
	B. animalis (ssp. lactis BB–12)
	B. adolescentis
	B. breve
Yeast and molds	*Saccharomyces cerevisiae*
	S. fragilis
	S. boulardii
	Torulopsis spp.[c]
	Aspergillus oryzae[c]
Bacillus	*Bacillus cereus*[c]
	B. toyoi[c]
	B. licheniformis[c]
	B. subtilis[c]
Others	*Enterococcus faecalis*[b]
	Enterococcus faecium[c]
	Streptococcus thermophilus

[a] Commercial names of specific strains are given in brackets.

[b] Mainly used in pharmaceutical preparations.

[c] Mainly used in animal husbandry.

Source: Ref. [13].

to the colon where they are metabolized and they promote the growth of specific advantageous microbiota (probiotics) by supplying food/energy, while simultaneously influencing the microbiota gene expression and enhance immunity [19, 20].

Additionally, fermentation of the prebiotics by the probiotics results in the production of beneficial byproducts, pathogen inhibitory substances (lactic acid, bacteriocin, etc.) restricting the growth of pathogens and stimulating of Immunoglobulin A (IgA) [21].

Prebiotics can be found in natural sources such as whole grains, asparagus, chicory, oats, soybeans, onions, bananas, garlic, honey, leeks, and artichokes and wheat. The raw vegetable material is also a key component of a high percentage of commercial prebiotics. Production is achieved via an enzymatic method, through the transglycosylation of monosaccharides or disaccharides, or the hydrolysis of complex polysaccharides [22, 23]. Several types of prebiotics and their sources are summarized in Table 2.2.

The only prebiotics for which sufficient data have been generated to allow an evaluation of their possible classification as functional food ingredients are the inulin-type fructans, which are linked by β (2–1) bonds that limit their digestion by upper intestinal enzymes, and fructooligosaccharides [18].

The utilization of prebiotics as food components has multiple advantages, since they improve sensory features and provide a more well-balanced nutritional composition [24]. Prebiotics are also often utilized as dietary fiber in tablets and in functional foods, particularly in entire ranges of dairy products and breads, as prebiotic ingredients enhance the viability of healthy intestinal bacteria [24].

The combination of a probiotic with a prebiotic to support its viability and activity has been termed a symbiotic [17, 25].

2.4 SYNBIOTIC

The term synbiotic refers indirectly to a *synergy*, and are defined as a combination of a probiotic and a prebiotic that beneficially affects the host by improving the survival, activity, and implantation of live microbiota and dietary supplements in the gastrointestinal tract, as well as selectively stimulating the growth and/or by activating the metabolism of indigenous

TABLE 2.2 Types and Sources of Prebiotics

Type of prebiotic	Chemical structure	Chemical composition	Natural source
Fructooligosaccharides (FOS)	(Fr)*n*-Gu	B-(2→1) linked fructosyl unit with terminal glucose. Chain length of 2–10 with average DP of 4	Asparagus, sugar beet, garlic, chicory, onion, Jerusalem artichoke, wheat, honey, banana, barley, tomato, and rye
Isomaltulose (IMO)	(Gu)*n*	4–7 units of α-(1–6) linked (6-O-α-D-glucopyranosyl-D-glucose)	Honey, sugarcane juice
Xylooligosaccharides (XOS)	Xy*n*	α-(1,6) linked galactose bonded via α-(1,3) to terminal sucrose β-(1–4) linked xylopyranose, with arabinofuranosyl, 4-Omethylglucuronic acid, acetyl, phenolics subsitutent at C2 or C3.	Bamboo shoots, fruits, vegetables, milk, honey, and wheat bran
Galactooligosaccharides (GOS)	(Ga)*n*-Gu	2–8 galactose units with terminal glucose having 1→4, 1→6 linkages	Human milk and cow's milk
Lactulose	Ga–Fr	4-O-β-D-galactopyranosyl-D-fructose	Lactose (milk)
Soybean oligosaccharide	Raffinose, stachyose	α-(1,6) linked galactose bonded via α-(1,3) to terminal sucrose	Soybean
Inulin	GuFr*n*	FOS having DP from 3–60. Fructosyl-glucose and fructosyl-fructose linked by β-(2–1) & β-(1–2) respectively	Chicory roots, onion, asparagus, antichoke

Gu: Glucose, Fr: fructose, Ga: Galactose, Xy: xylose, *n*: number of residues. DP: Degree of polymerization, RT: Room temperature.

Source: Ref. [24].

bifidobacteria and lactobacilli [9, 17, 18, 26]. Figure 2.1 purported the mechanism of action of probiotics, prebiotics, and their synergism.

Recent *in vitro* studies have demonstrated that synbiotic were more effective than prebiotics or probiotics in modulating the gut microflora [25]. Several authors have recommended that the ingestion of 10 g/day of galactooligosaccharides is adequate to cause a bifidogenic effect. Daily ingestion of 2.5 g prebiotics/day is sufficient to increase fecal bifidobacteria levels [27], however for xylooligosaccharides, 2 g/day is considered enough to incur bifidogenic effect [28], the fructooligosaccharides (FOS) intake necessary to perform as a bifidogenic stimulus is between 2 and 10 g/day in adults. Nevertheless, at least 4 g FOS/day would be required to raise the bifidobacteria levels in the human gut [29]. The daily amount needed for isomaltooligosaccharides is in the range of 8–10 g [30].

Fermented milk (yogurt and kefir) are true synbiotic products, that is, functional foods, since they contribute to restore the normal bacterial microflora also supplying the food it needs to survive. However not all these products promote symbiosis, the best synbiotic combinations

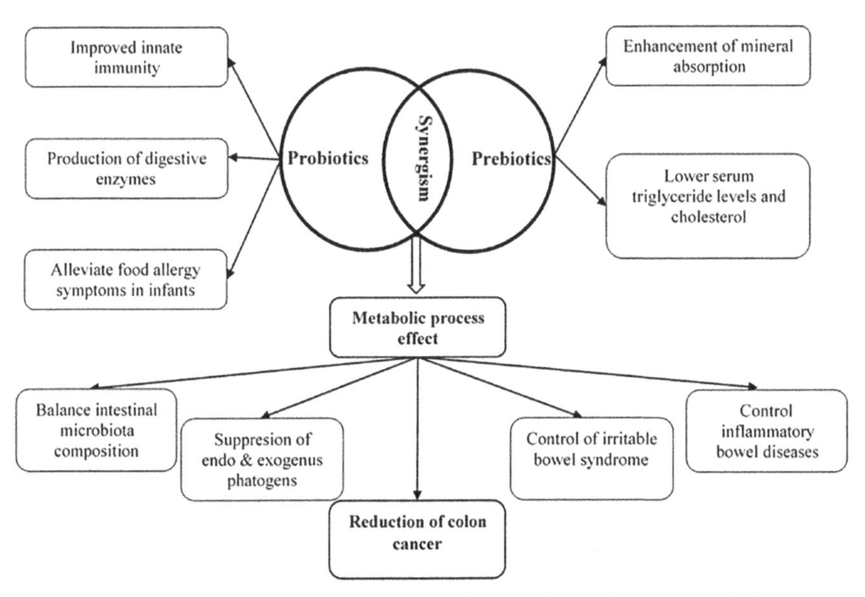

FIGURE 2.1 Purported mechanism of action of probiotics, prebiotics, and their synergism.

currently available include bifidobacteria and FOS, Lactobacillus GG and inulins, and bifidobactera and lactobacilli with FOS or inulins [8].

One of the principal benefits of synbiotic is believed to be increased persistence of the probiotic in the gastrointestinal tract. A synbiotic

TABLE 2.3 Some Nondairy Probiotic Products Recently Developed

Category	Substrates	Probiotic microorganism	Products
Fruit and vegetable based	Cassava	*Lactobacillus plantarum, L. brevis, L. fermentum, Leucostoc mesenteroides*	Agbelima
	Cabbage	*L. mesenteroides, Lactococcus lactis*, LAB	Sauerkraut
	Ginger juice	*Lactobacillus, Leuconostoc, Bacillus, Staphylococcus, Candida,* and *Saccharomyces*	Ginger beer
	Vegetables	*L. plantarum, L. curvatus, L. brevis, L. sake, L. mesenteroides*	Kimchi
Cereal based	Legumes	*LAB*	Adai
	Cercals	*L. plantarum, L. brevis, L. rhamnosus, L. fermentum, L. mesenteroides* subsp. *dextranium*	Boza
	Maize	*LAB*	Pozol
	Sorghum	*Lactobacillus* sp., *L. brevis*	Kisra
	Maize	*LAB*	Ilambazi lokubilisa
	Pearl millet	*LAB*	Bensaalga
	Maize	*L. casei, L. lactis, L. plantarum, L. brevis, L. acidophilus, L. fermentum, L. casei*, yeast	Kenkey
	Rice and Bengal gram	*L. mesenteroides, L. fermentum, S. cerevisiae*	Dosa
	Millet	*L. fermentum, L. Salivarius*	Koko
	Cereals	*L. casei, S. cerevisiae, L. mesenteroides*	Beet kvass
	Maize and finger millet	*L. plantarum, L. brevis, L. fermentum, L. cellobiosus, Pediococcus pentosaceus, Weissella confusa, Issatchenkia orientalis, S. cerevisiae, Candida pelliculosa and Candida tropicalis*	Togwa
	Sorghum or millet	*Lactobacillus, Lactococcus, Leuconostoc, Enterococcus*, and *Streptococcus, L. brevis*	Bushera
Soy based	Soybean	*LAB, L. plantarum*	Tempeh
	Soybean, wheat	*LAB*	Kecap

Source: Ref. [3].

preparation of *Lactobacillus acidophilus* (probiotic strain 74–2) and FOS has been studied in an *in vitro* model of the human gut [31].

The consumption of beverages and foods that contain probiotic microorganisms is a growing worldwide trend [32]. Even though fermented dairy products are generally good matrices for delivery of probiotic to humans, other foods have been examined for their potential as probiotic carriers. Mayonnaise, soymilk, meats, baby foods, ice cream, fruit drinks, vegetable drinks, and many others have already been proposed, as shown in Table 2.3 [3].

This demonstrates how scientists are overcoming some of the obstacles associated with probiotic survival. Currently, the most successful synbiotic products have been dairy foods such as yogurts. Fruit/vegetable juices and cheese are future synbiotic candidates [19].

2.5 EXPECTED THERAPEUTIC USES

Most health effects attributed to synbiotic are related, directly or indirectly, i.e., mediated by the immune system, to the gastrointestinal tract. This is not only due to the fact that probiotics and prebiotics in food or therapeutically used synbiotics are applied normally via the oral route [31]. The mechanisms and the efficacy of a synbiotic effect often depend on interactions with the specific microflora of the host or immunocompetent cells of the intestinal mucosa. The gut (or the gut-associated lymphoid system (GALT), respectively), is the largest immunologically competent organ in the body, and maturation and optimal development of the immune system since birth depends on the development and composition of the indigenous microflora [13].

The commercial interest in functional foods that contain probiotics, is paralleled by increasing study of their role in the digestive tract. To date, interest in probiotic containing foods has centered around three health propositions: (i) improving general gut health, (ii) improving body natural defenses, and (iii) lowering blood cholesterol [33]. Table 2.4 shows the probiotic, prebiotic, and synbiotics products currently on the market.

TABLE 2.4 Some Commercially Marketed Formulations of Probiotics, Prebiotics, and Synbiotics

Product name	Company	Country	Probiotic bacteria present	Description and properties of the product
Kyo-Dophilus	Wakunaga	USA	*Lactobacillus acidophilus* *Bifidobacterium bifidum* *B.longum* *L. acidophilus*	Contains human strains of bacteria (each capsules), capsule contains about 1.5 billion live cells.
Lactiflora	Dr. Reddy's	India	*L.acidophillus*	This product improves the patient's condition by suppressing or killing harmful bacterial growth, as well as the prevention and treatment of pediatric loose motions.
Kyo-Dophilus tablets	Wakunaga	USA	*L. acidophilus*	Stable preparation not requiring refrigeration, completely vegetarian, dairy and sugar-free, chewable, tasty and convenient for travel.
Becelac (capsules)	Dr. Reddy's	India	*L.acidophilus*, thiamine, vitamin B–2, vitamin B–6, vitamin B–12, vitamin C, acid folic, niaciamide, calcium panthonate.	This product is a medicine that is used for the treatment of diarrhea, treatment of megaloblastic anemias due to a deficiency of folic acid, treatment of anemias of nutritional origin, pregnancy, infancy, or childhood, vitamin B–12 deficiency, pernicious anemia, supplements, and other conditions.
Acidophilas®	Wakunaga	USA	*L. acidophilus*, lipase, protease, amylase, and lactase enzymes	Enzymes assist in the breakdown of fats, proteins, and carbohydrates, lactase assists in the digestion of milk sugar lactose in lactose intolerance.
Lactisyn	Franco Indian Pharmaceuticals Ltd.	India	*L.lactis* *L. acidophilus* *S. thermophilus* *S.lactis*	Restoration of lactobacillus flora after anti-infective/chemotherapeutic treatment; atrophic vaginitis due to estrogen deficiency, as co-medication to hormone replacement therapy; vaginal discharge of unknown origin or mild to moderate cases of bacterial vaginosis & candidiasis.

TABLE 2.4 *(Continued)*

Product name	Company	Country	Probiotic bacteria present	Description and properties of the product
Probiata® tablets	Wakunaga	USA	An internationally accepted strain of *L. plantarum*	Bacteria pre-adapted for growth in human intestine. Each tablet contains one billion live cells per tablet, preparation no requires refrigeration and is heat resistant. Survives when taken with food and resistant to acid of stomach (pH 3–4).
Flora Grow 90	Arise & Shine Herbal products Inc.	USA	*B. infantis* *B. Longum* *B. bifidum*	Generate a pH between 6.5 and 7 instead of pH 4.5, as produced by *Lactobacillus acidophilus*.
Bifa 15	Eden Foods	USA	*B. longum* *L.acidophilus* *L.gasseri*	Contains microencapsulated *B. longum* designed to get past the stomach acids and reach the colon. Produces a very hostile environment for HIV, *Candida*, and other infections of the large intestine.
TH1 Probiotics	Jarrow Formulas	USA	*L. casei* *L. plantarum* *Saccharomyces boulardi*	Contains one billion *Bifidobacterium longum* per serving along with ½ billion *Saccharomyces boulardi*, both *L. casei* and *L. plantarum* are heat treated so as to be safe for use in severely immune compromised persons who have leaky gut.
Replenish	Innercleanse 2000		*L.plantarum* *L.bifidus* *L.bulgaricus* *L.rhamnosus* *L.casei* *L.brevis* *Fructooligosaccharides* Stabilized *Bifidobacteria*	Causes an increase in production of beneficial short-chain fatty acids, reduction of serum cholesterol and blood pressure, improved liver function and improved elimination of toxic compounds. FOS helps promote the growth of friendly bacteria. While simultaneously reducing the colonies of detrimental bacteria.

TABLE 2.4 *(Continued)*

Product name	Company	Country	Probiotic bacteria present	Description and properties of the product
VSL#3	Alfasigma	Italian	5×10^{11} cells/g of strains: *B. breve* *B. longum* *B. infantis* *L.acidophilus* *L. plantarum* *L. paracasei* *L. delbrueckii (bulgaricus)* *Streptococcus thermophilus*	VSL#3 unique synergistic composition, with 8 diverse strains of bacteria, has significant advantages in terms of metabolic and functional activities. Colonize the intestine and help in the remission of ulcerative colitis in the patients intolerant of allergic to 5-aminosalicylic acid.
Subalin Forte	Biofarma	Ukrainian	Live bacteria *Bacillus subtilis*	Subalin Forte helps to normalize intestinal microflora, prevent, and correct dysbiosis, reduce endogenous intoxication and increase resistance to viral infections.
Colinfant, mutaflor	Dyntec	Czech Republic	Non-enteropathogenic *Escherichia coli cryodesiccata*	Live oral vaccines, which colonize the intestines of full-term and pre-term infants and establish themselves as resident strains. Stimulate antibody production in the gut, saliva, and serum of colonized infants. Decrease the presence of pathogenic bacterial strains in the intestines as well as on other mucosal surfaces of the body.
Symbiotik capsules	• Cadila Pharma • Le Sante Cadila Pharma	India	*L. sporogenes* Antibiotics (amoxicillin and cloxacillin)	Each capsule contains 250 mg amoxicillin, cloxacillin and 60 million spores of *Lactobacillus sporogenes*, useful for respiratory tract infections, ENT infections, skin, and soft tissue infections, urinary.

TABLE 2.4 *(Continued)*

Product name	Company	Country	Probiotic bacteria present	Description and properties of the product
Symbiotik-P tablets	Cadila Pharma	India	*Lactobacillus sporogenes* Antibiotics (amoxicillin and cloxacillin)	Symbiotik-P is used for bacterial infections of skin, bacterial infections, the treatment of beta-hemolytic streptococcal infections, the treatment of pneumococcal infections, and the treatment of staphylococcal infections, diarrhea, and other conditions.
Primal Defense™ capsules	Garden of life	USA	HSO™ *L. acidophilus* *L. rhamnosus* *L. salivarius* *L. plantarum* *L. paracasei* *L. casei* *L. brevis* *Bifidobacterium bifidum* *Bifidobacterium breve* *Bifidobacterium lactis* *Bifidobacterium longum* *Bacillus subtilis* *Saccharomyces boulardii*	Primal Defense is the only probiotic formula that contains a unique whole food blend with Homeostatic® Soil Organisms (HSOs). Promotes healthy digestive balance, supports normal absorption and assimilation of nutrients, helps support digestive comfort and helps maintain a healthy immune system.
Culturelle	i-Healt Inc	USA	*L. GG* Inulin (chicory)	Keep digestive system in balance, reduces the likelihood of occasional diarrhea, gas, and bloating

2.6 CONCLUSION

Probiotics are live nonpathogenic microorganisms which have a beneficial effect on the health of the host. They are present in the gastrointestinal tract without causing any adverse effects. Probiotics can be used for several clinical applications, e.g., antibiotic-induced diarrhea, irritable bowel syndrome and inflammatory bowel disease. Prebiotics are known to be a non-digestible food ingredient, and they have a significant effect on human health its action is based by the fermentation of carbohydrates, which stimulate, preferentially, the growth of probiotic bacteria (bifidobacteria and lactic acid bacteria), thus enhancing the gastrointestinal and immune systems. It has also been demonstrated, the prebiotics increases the absorption of calcium and magnesium, influence blood glucose levels and improves plasma lipids. The properties of prebiotics have greater possibilities for incorporation into a wide range of common foodstuffs. There is sufficient scientific evidence to suggest the beneficial role of the synbiotics in the maintenance of good health and the prevention of infectious and metabolic diseases. However, there are significant gaps between what research has shown to be effective for some products, and what is claimed in the marketplace. The health professional is in an ideal position to guide the consumer towards appropriate prophylactic and therapeutic uses of synbiotics that deliver the desired beneficial health effects. Once it is achieved, it should be possible to develop specific dietary recommendations for diseases prevention.

KEYWORDS

- **food processing**
- **functional food health wellness**
- **prebiotics**
- **probiotics**
- **synbiotic**
- **synergism**

REFERENCES

1. Morais, E. C., Morais, A. R., Cruz, A. G., & Bolini, H. M. A., (2014). Development of chocolate dairy dessert with the addition of prebiotics and replacement of sucrose with different high-intensity sweeteners, *J. Dairy. Sci.*, *97*, 2600–2609.
2. Siró, I., Kápolna, E., Kápolna, B., & Lugasi, A., (2008). Functional food. Product development, marketing and consumer acceptance-A review, *Appetite.*, *51*, 456–467.
3. Granato, D., Branco, G. F., Nazzaro, F., Cruz, A. G., & Faria, J. A. F., (2010). Functional foods and nondairy probiotic food development: Trends, concepts, and products, *Compr. Rev. Food Sci. Food Saf.*, *9*, 292–302.
4. Perricone, M., Bevilacqua, A., Altieri, C., Sinigaglia, M., & Corbo, M., (2015). Challenges for the production of probiotic fruit juices, *Beverages*, *1*, 95.
5. Cencic, A., & Chingwaru, W., (2010). The role of functional foods, nutraceuticals, and food supplements in intestinal health, *Nutrients*, *2*, 611–625.
6. Macfarlane, S., Macfarlane, G. T., & Cummings, J. H., (2006). Review article: Prebiotics in the gastrointestinal tract, *Aliment. Pharmacol. Ther.*, *24*, 701–714.
7. Vijaya, K. S. G., Singh, S. K., Goyal, P., Dilbaghi, N., & Mishra, D. N., (2005). Beneficial effects of probiotics and prebiotics on human health, *Pharmazie*, *60*, 163–171.
8. Iannitti, T., & Palmieri, B., (2010). Therapeutical use of probiotic formulations in clinical practice, *Clin. Nutr.*, *29*, 701–725.
9. Tuohy, K. M., Probert, H. M., Smejkal, C. W., & Gibson, G. R., (2003). Using probiotics and prebiotics to improve gut health, *Drug Discov. Today*, *8*, 692–700.
10. Tripathi, M. K., & Giri, S. K., (2014). Probiotic functional foods: Survival of probiotics during processing and storage, *J. Funct. Foods.*, *9*, 225–241.
11. Gibson, G. R., (1998). Dietary modulation of the human gut microflora using prebiotics, *Br. J. Nutr.*, *80*, S209–212.
12. Bielecka, M., Biedrzycka, E., & Majkowska, A., (2002). Selection of probiotics and prebiotics for synbiotics and confirmation of their *in vivo* effectiveness, *Food Res. Int.*, *35*, 125–131.
13. De Vrese, M., & Schrezenmeir, J., (2008). Probiotics, prebiotics, and synbiotics, In: Stahl, U., Donalies, U. E. B., & Nevoigt, E., (eds.), *Adv. Biochem. Eng./Biotechnol.* (pp. 1–66). Springer, Berlin Heidelberg.
14. Kaur, I. P., Chopra, K., & Saini, A., (2002). Probiotics: Potential pharmaceutical applications, *Eur. J. Pharm. Sci.*, *15*, 1–9.
15. Gibson, G. R., Hutkins, R., Sanders, M. E., Prescott, S. L., Reimer, R. A., Salminen, S. J. et al., (2017). Expert consensus document: The International Scientific Association for Probiotics and Prebiotics (ISAPP) consensus statement on the definition and scope of prebiotics, *Nat. Rev. Gastroenterol Hepatol.*, *14*, 491–502.
16. Balasubramaniem, A. K., Nagarajan, K. V., & Paramasamy, G., (2001). Optimization of media for β-fructofuranosidase production by *Aspergillus niger* in submerged and solid state fermentation, *Process Biochem.*, *36*, 1241–1247.
17. Gibson, G. R., & Roberfroid, M. B., (1995). Dietary modulation of the human colonic microbiota: Introducing the concept of prebiotics, *J. Nutr.*, *125*, 1401–1412.
18. Quigley, E. M. M., (2010). Prebiotics and probiotics, modifying and mining the microbiota, *Pharmacol. Res.*, *61*, 213–218.

19. Scourboutakos, M., (2010). 1 + 1 = 3...Synbiotics: Combining the power of pre- and probiotics, *J. Food Sci. Educ.*, *9*, 36–37.
20. Oliveira, R. P. d. S., Perego, P., Oliveira, M. N. D., & Converti, A., (2011). Effect of inulin as prebiotic and synbiotic interactions between probiotics to improve fermented milk firmness, *J. Food Eng.*, *107*, 36–40.
21. Vieira, A. T., Teixeira, M. M., & Martins, F. S., (2013). The role of probiotics and prebiotics in inducing gut immunity, *Front Immunol.*, *4*, 445.
22. Szkaradkiewicz, A. K., & Karpiński, T. M., (2013). Probiotics and prebiotics, *Eur. J. Biol. Res.*, *3*, 6.
23. Combrinck, Y., & Schellack, N., (2015). Probiotics and prebiotics. A review, *Prof. Nurs. Tod.*, *19*, 16–20.
24. Al-Sheraji, S. H., Ismail, A., Manap, M. Y., Mustafa, S., Yusof, R. M., & Hassan, F. A., (2013). Prebiotics as functional foods: A review, *J. Funct. Foods.*, *5*, 1542–1553.
25. Davis, C. D., & Milner, J. A., (2009). Gastrointestinal microflora, food components and colon cancer prevention, *J. Nutr. Biochem.*, *20*, 743–752.
26. Saulnier, D. M. A., Gibson, G. R., & Kolida, S., (2008). *In vitro* effects of selected synbiotics on the human faecal microbiota composition, *FEMS Microbiol. Ecol.*, *66*, 516–527.
27. Gibson, G. R., (1999). Dietary modulation of the human gut microflora using the prebiotics oligofructose and inulin, *J. Nutr.*, *129*, 1438–1441.
28. Sako, T., Matsumoto, K., & Tanaka, R., (1999). Recent progress on research and applications of non-digestible galacto-oligosaccharides, *Int. Dairy J.*, *9*, 69–80.
29. Manning, T. S., & Gibson, G. R., (2004). Prebiotics, *Best Pract. Res., Clin. Gastroenterol.*, *18*, 287–298.
30. Goulas, A. K., Fisher, D. A., Grimble, G. K., Grandison, A. S., & Rastall, R. A., (2004). Synthesis of isomaltooligosaccharides and oligodextrans by the combined use of dextransucrase and dextranase, *Enzyme Microb. Technol.*, *35*, 327–338.
31. Pandey, K. R., Naik, S. R., & Vakil, B. V., (2015). Probiotics, prebiotics and synbiotics: a review, *Journal of Food Science and Technology*, *52*, 7577–7587.
32. Verbeke, W., (2005). Consumer acceptance of functional foods: socio-demographic, cognitive and attitudinal determinants, *Food Qual. Prefer.*, *16*, 45–57.
33. Parvez, S., Malik, K. A., Ah Kang, S., & Kim, H. Y., (2006). Probiotics and their fermented food products are beneficial for health, *J. Appl. Microbiol.*, *100*, 1171–1185.

CHAPTER 3

RECENT TRENDS IN EXTRACTION, PURIFICATION, AND CHARACTERIZATION OF BIOACTIVE PEPTIDES

MARÍA L. CARRILLO-INUNGARAY,[1] JESÚS A. MARTÍNEZ-SALAS,[1] MARIELA MICHEL,[2] JORGE E. WONG-PAZ,[3] DIANA B. MUÑIZ-MÁRQUEZ,[3] and PEDRO AGUILAR-ZÁRATE[3]

[1] *Food Research Laboratory, Autonomous University of San Luis Potosí, 79060, Ciudad Valles, San Luis Potosí, México*

[2] *Food Research Department, School of Chemistry, Autonomous University of Coahuila, 25280, Saltillo, Coahuila, Mexico*

[3] *Department of Engineering, Tecnológico Nacional de México, Campus Ciudad Valles, 79010, Ciudad Valles, San Luis Potosí, Mexico, Tel.: +52 481 3812044, E-mail: pedro.aguilar@tecvalles.mx*

ABSTRACT

Bioactive peptides are protein fragments with a positive impact on the human health. Peptides have shown several biological properties such as antimicrobial, antiviral, antitumor, among other activities and they are produced by almost all species. These peptides have attracted the interest of researchers and consumers due to the high potential use in functional foods, cosmetic and pharmaceutical industries, and other dietary interventions for disease control and health promotion. However, there are important topics of concern prior to the application of bioactive peptides. For that reason, this chapter reviews the recent trends in extraction, purification, and characterization of bioactive peptides. It includes the strategies and advances in extraction technologies of bioactive peptides from different sources. Also, the purification methodologies employed in the last five years are discussed. Even more, it is presented the recent trends in the characterization of bioactive peptides, including bioactivities, toxicology, and the aminoacidic sequencing.

3.1 INTRODUCTION

Bioactive peptides are short sequences of amino acids with a low molecular weight that may be generated from sources such as milk, egg, fish, plants, etc. Generally, the peptides are inactive within the original sequence of the protein [1]. Bioactive peptides have demonstrated to possess activities such as antimicrobial, antioxidant, anticancer, immunomodulatory, among others. Also, are effective in the treatment of diseases as cancer, diabetes, and obesity [2–5]. On the human body, liberated bioactive peptides act as regulatory compounds similar to hormones.

The bioactivities of peptides depend on their amino acid composition and sequence. Notwithstanding, it is known some peptides are multifunctional and can present multiple activities [6] depending on multiple factors, for example, the source and the extraction method. In recent years, the scientific research on the production, purification, and characterization of bioactive peptides from different sources have increased.

Research works report methods for the extraction of bioactive peptides that includes the use of typical methods such as maceration or Soxhlet, alternative technologies as microwave and ultrasound assisted extraction, and the enzymatic extraction [7, 8]. The relatively low cost of used raw material, the environmentally safe and economic process make the generation of bioactive peptides an attractive option.

The downstream process in the production of bioactive peptides involves the concentration and purification for further application. Methodologies such as precipitation and ultrafiltration (UF) have been used for the concentration of bioactive peptides as an initial step in the purification [9–11]. Chromatography has been used for the concentration (fractionation), purification, and characterization of peptides by obtaining the aminoacidic sequences (coupled with mass spectrometry) [12–15].

The application of bioactive peptides in different industries and overall, because of their potential benefits on human health are the main causes for the development of research works [16]. Depending on the source of bioactive peptides, the nature of protein and the equipment available on laboratory, researchers have carried out many strategies for the extraction, purification, and characterization of bioactive peptides. For that reason, the present document reviews the recent trends in the extraction methods of bioactive peptides from different sources. Also, presents some of the most used methods for the purification of bioactive peptides reported in the last five years. Besides, discusses the characterization of bioactive peptides by their biological activities and presents information related to the tools recently used for the mapping and *in silico* analysis of peptides.

3.2 SOURCES OF BIOACTIVE PEPTIDES

Bioactive peptides have been isolated from different sources, both animal and plant; among them can be mentioned: casein, muscle of chicken and fish among others and vegetable origin have been isolated from wheat gluten, soybean [17], sunflower, spinach, mung beans [18], dairy products [19], egg, meat [20, 21], and marine products [22].

Although a peptide can be easily synthesized by a chemical method, it is almost impossible to resynthesize a glycopeptide by a chemical method not only because the chemical synthesis of oligosaccharides involves complicated steps, but also because of the difficulty in binding it with the peptide [23]. Among the strategies for glycopeptide synthesis include the chemical addition of oligosaccharides to peptides by reductive amination, solid-phase synthesis using glycol-aminoacids as building blocks, and enzymatic synthesis using peptidases and glycosyltransferases.

The bioactive peptides have been found hidden in salivary parent proteins such as statherin, histatin 3, histatin 1, three proline-rich proteins (PRPs) [P-B1 (alias SMR3A), P-B (SMR3B), and BPLP (basic proline-rich lacrimal protein)], and mucin 7 (MG2), all of which are encoded by a gene cluster on chromosome 4q13.3 [24].

3.2.1 PLANTS

Salas [25] reviewed the biological and antimicrobial activity of epitopes obtained from plants. Full isolation of plant AMP has been attained in some cases. It is the case of lunatusin a peptide with a molecular mass of 7 kDa purified from Chinese lima bean (*Phaseolus lunatus* L.). Lunatusin exerted antibacterial action on *Bacillus megaterium*, *Bacillus subtilis*, *Proteus vulgaris*, and *Mycobacterium phlei*. The peptide also displays antifungal activity towards *Fusarium oxysporum*, *Mycosphaerella arachidicola*, and *Botrytis cinerea*. Interestingly, the antifungal activity was retained after incubation with trypsin [26].

3.2.2 ANIMALS

Ambigaipalan and Shahidi [27] identified three potential bioactive peptides from shrimp shell discard protein hydrolysate fractions with the highest inhibition to Angiotensin Converting Enzyme (ACE). The proteins were further subjected to gel filtration, mass spectrometry, and peptide sequencing. Three potential ACE inhibitory bioactive peptides were identified form shrimp shell discard for the first time. Identification of bioactive peptide from shrimp shell discard protein hydrolysates was subjected to gel filtration chromatography. An electrophoresis was conducted to confirm the protein in the fractions. The fractions were further subjected to analysis of ACE inhibitory activity and the fractions exhibiting highest ACE inhibition were further purified with C18 Zip tips. The amino acid composition of raw shrimp shell was calculated with respect to the area of internal standard (norleucine).

Abd El-Fattah [28] obtained bioactive peptides from milk. In this study, milk was hydrolyzed using protease (*Asperigillus oryzae*), trypsin, pepsin, or papain at concentrations of 0.001, 0.005, or 0.01g/100 g milk for 30 or 60 min to produce angiotensin-converting enzyme inhibitory and antioxidant peptides. Results showed that the proteolysis, antioxidant, and ACE-I activity gradually increased with the increase in the enzyme concentration and hydrolysis time. The protease-treated milk had the highest proteolytic and ACE-I activity, while the papain-treated milk had the lowest. The papain-treated milk exhibited the greatest Fe^{2+} chelating activity. The use of trypsin at a concentration of 0.001g/100 g milk for 60 min produced ACE-I and antioxidant activity without changes in the technological properties of milk. Also, from breast milk has been derived bioactive peptides proteins, such as caseins, α-lactalbumin, and lactoferrin, during gastrointestinal digestion.

On the other hand, Nimalaratne [29] examined the ability of several proteases for their ability to release antioxidant peptides from hen egg white, and protease P was selected based on the antioxidant activity and the digestion yield of the crude protein hydrolysate. A combination of several purification steps including ultrafiltration with low molecular weight cut-off membranes, cation exchange chromatography and reversed phase high-performance liquid chromatography was used to purify 'protease P egg white hydrolysate.' Sixteen antioxidant peptides, which were derived from ovalbumin, ovotransferrin, and cystatin, were isolated from the most active fractions.

Mohanty [30] reported that peptides derived of the milk exert a number of health beneficial activities, even upon oral administration, and are considered as potent drugs with well-defined pharmacological residues and also used to formulate health-enhancing nutraceuticals for its impact on major body systems including the digestive, nervous, endocrine, cardiovascular, diabetes type II, obesity, and immune systems.

3.2.3 MICROBIAL

Yamamoto [23] described that the chemo-enzymatic synthesis of a glycopeptide, which involves the chemical synthesis of Nacetylglucosaminyl peptide and the enzymatic transfer of oligosaccharide. Endo-P-N-acetylglucosaminidase (endo+GlcNAc-ase, EC 3.2.1.96) is a unique endoglycosidase that hydrolyses NJ'-diacetylchitobiosyl linkages in oligosaccharides bound to asparaginyl residues of various glycoproteins and glycopeptides, and leaves one N-acetylglucosamine (GlcNAc) residue on the protein and peptide moieties. He found a novel endo+-GlcNAc-ase in the culture medium of *Mucor hiemalis* isolated from soil, which could cleave not only the high mannose and hybrid-type asparagine-linked oligosaccharides but also the complex-type oligosaccharide (7, S), unlike other endo-P-GlcNAc-ases that can act on only the high mannose and hybrid-type oligosaccharides. We named this novel enzyme Endo-M after its source. This enzyme showed transglycosylation activity and could transfer the oligosaccharides from glycopeptides to suitable acceptors with a GlcNAc residue during hydrolysis of the glycopeptide.

Bioactive peptides can be produced by enzymatic hydrolysis using proteolytic enzymes of the digestive system or microbial origin, microbial fermentation and food processing. Using single or multiple specific or nonspecific proteases, is the most common way for bioactive peptide production because it requires a shorter time to obtain a similar degree of hydrolysis as well as better control of the hydrolysis to obtain more consistent molecular weight profiles and peptide composition. These processes are especially used in food and pharmaceutical industries by several enzymes such as Pepsin, Bromelain, Trypsin, Chymotrypsin, and Papain under their respective optimal pH and temperature [31].

3.3 TECHNOLOGIES FOR THE EXTRACTION OF PEPTIDES

In recent years the use of bioactive peptides of various plant materials for commercial sectors has been increasing, becoming more and more required, beginning with health studies, and ending with industrialization, with that began to develop different extraction techniques to acquire bioactive compounds [32]. Each technique having different objectives, either to extract specific compounds from the plant sample, increase the selectivity of methods during extraction, increase the sensitivity and concentration of a certain compound. Most extraction techniques use heat, solvent, or mixture. Thus, we have a classification in which the types of peptide extractions are included – conventional technologies, alternative technologies and green technologies (enzyme-assisted extraction) [33].

3.3.1 CONVENTIONAL TECHNOLOGIES

Conventional extraction technologies are particularly constituted by methods of aqueous extraction, Soxhlet extraction, solid-liquid extraction that is also referred to as maceration. Liquid extraction is a high-performance technique, the technique being mostly used for the identification and purification of peptides having as characteristic during the extraction of a high yield and a recovery of material are obtained. This technique has a very effective approach used for the extraction of coconut, soybean, and maize germ, obtaining quality proteins. The different factors influencing the efficiency of aqueous extraction is the solid-water interface, types of salts and their concentrations pH, temperature, as well as the time the entire process lasted together [32].

Solid-liquid maceration or extraction techniques have been used for the home preparation of a long-term tonic, being a widely used and low-cost way to obtain bioactive oils and compounds, crush the leaves of plant materials, mix, and aggregate Solvents suitable for each type of extraction. Achievement of the essential part of the sample that was treated [33]. The extraction technique is widely used in the area of industrial pharmacy particularly in the perfume and plaguicide industries, to have a recovery of active plant products. The disadvantages of using this technique is the use of heat, since solvents are used during extraction, as the heating act as steam to release bioactive compounds from plant tissues. Molecular affinity is very important because of the solvent-solute relationship, since at high temperature some volatile compounds are lost. This extraction technique is often useful when putting together a first fraction strategy to remove residues and thus obtain an abundant peptide concentration [34].

For decades, conventional methods, such as the Soxhlet extraction technique, have been used in the environmental area, as well as for different purposes, characterized by exceeding extraction efficiency. Characterized as the universal extraction, by the time in which it has been used [35]. The disadvantages of this method of extraction are time, despite being a standard technique for obtaining active samples, has been replaced by alternative techniques, which guarantee the same performance in a shorter time [36]. The technique has been compared with alternative extraction methods, mainly with ultrasonic assisted extraction, since they achieve similar behaviors during their extraction, being used to extract oils. The pressure, and temperature can significantly influence the extraction [37].

Another extraction technique is the ultrafiltration membrane system, which separates the proteins and the macromolecule having bioactivities and specific functional properties of the extract sample material. These membranes are used mainly in vegetable material such as rapeseed, soybean, oats, canola, as well as fruit juices, wines, and beers, being of great interest in recent years in commercial industries, for demonstrating that recovering at the time extraction of high concentrations of proteins and peptides [32].

Several authors point out that membrane filtration may have an incomplete removal of proteins, being a modern extraction technique. It has the advantage of high heat output, a selective separation during the filtration does not require the application of heat or high temperatures for the extraction, nor does it occupy external additives or substances. The drawbacks of modern extraction technologies are the high cost of membranes, and runs have a very short shelf life due to membrane saturation and fouling [34].

3.3.2 ALTERNATIVE TECHNOLOGIES

Alternative technologies are used for the extraction of active compounds from plant materials. The plant materials have the characteristic of being sensitive to light, at elevated temperatures, and to large amounts of solvents. These extraction methods vary somewhat with the performance they can provide compared to conventional technologies [8]. The techniques of extraction into this classification of technologies, are ultrasound-assisted extraction and microwave-assisted extraction [38].

The ultrasound-assisted is a novel technology extensively reported for the extraction of bioactive compounds from natural products [8] used to extract proteins from diverse sources, on the effect of ultrasound and the enzymatic treatment on soluble proteins and allergenic proteins [13, 39].

Compared to conventional extraction methods, it requires less time and less solvent, which can damage and pollute the environment. Variables in terms of time, type of solvent, liquid-solid ratio, and ultrasound power have much to do with improving extraction efficiency, since it is very different from extracting active compounds than oils from various plant materials [35].

Different authors affirm that the extraction by ultrasound increases the solubility of the protein and improves the cleavage of the peptide bonds, thus loosening the structure of the nucleus. The rate of extraction of peptides depends on the method of extraction, when things do not fully develop they retain stable and compact structures and are difficult to extract [39].

It has been mentioned previously that ultrasound which is a robust, green, and rapid technology suitable for scale-up, which can enhance the effectiveness of protein digestion, extraction, production, and drug delivery of bioactive peptides principally acts for by generating bubble cavitation in the biological matrix [8].

Compared with conventional procedures, alternative methods ending up as interesting techniques, used for extraction as novel and optimal methods for their use [37]. Another alternative technology is the microwave energy for extraction. It has been cataloged as a green method for the extraction of bioactive compounds and oils, widely used in the food industry, surpassing the performance of conventional techniques [40]. The most common drawbacks in this technique could be to damage the sample, since if there is no management in time, and intensity, it may not obtain extract or even worse leave the sample unusable vegetable [8]. A more modern use of the microwave extraction are the processes where they are using solvents, in which it shows a greater amount of extraction of matter to be extracted, having a 30% more yield in less time [7].

3.3.3 GREEN TECHNOLOGIES (ENZYME-ASSISTED EXTRACTION)

The demand in the last years for exploiting our natural resources has been intensified, thus having the need to occupy the active compounds derived from plants. To do this, it is necessary to improve the processes of releasing the bioactive from plant cells by either altering and extracting using specific enzymes. Green technologies supported by the biotechnology have not been fully exploited by industries. Enzyme-assisted extraction methods have been used because of the need for extraction of ecological means involving the known and established technologies, having relatively a favorable amount in the yield and extraction of extractive competences of said means [41].

The enzyme-assisted extraction process is a feasible alternative to conventional extraction technologies. It is a promising method for the simultaneous extraction of oil and proteins from plant materials, foods, and animal sources [42]. In investigations for the extraction of free amino acids and peptides, it was found to be of great yield in fungi and the most common are shiitake (*Lentinus edodes*), oyster (*Pleurotus ostreatus*), tea tree (*Agrocybe aegerita*) Brown and portobello champignons (*Agaricus bisporus*) [43].

The study of the application of enzymes for the methods of extraction of bioactive compounds has an enormous incidence in diverse industries. The yields on the basis of the obtaining are influenced by diverse factors such as the extraction temperature, which is handled at 55°C, the management of pH, the enzyme concentration or exposed amount, the extraction time, etcetera [41, 43, 44].

Some the enzymes most used for enzyme-assisted extraction are β-glucanase, flavorzyme, α-amylase, viscozyme, cellulase, protease, and pectin because they have a high efficiency of both at the time of extraction, up to 20-fold time than conventional extraction [45]. The enzymes that are used in this technique have the following optimal conditions for the extraction, water as an extraction solvent, initial pH = 7, the enzyme concentration of 5% v/w and maintain a temperature of 50°C and incubation time of 1 h [43].

The use of cellulase and protease makes a substantial difference in the performance of green extraction processes. Carrying out with a suitable mixture of solvents, and material, the process reaches 39% to 90% extraction yield compared to other extraction methods. The benefit is high when using these techniques for the extraction of bioactive extractable matter [46]. The application and effects of the enzymes α-amylase, viscozyme, cellulase, protease, and pectinase are very noticeable in foods such as ginger. Since similar foods have a yield difference of 13% to 30% for properties, and pre-treatment made with the solvents involved in the extraction. Significant amounts of proteins obtained by this extraction process are used in the food industries, in the manufacture of soft drinks and up to a certain percentage in perfumery. The green extraction technique with its enzymatic pre-treatment is considered an efficient, safe way of releasing the limited compounds and increasing the yield with their respective specific enzymes [33].

3.4 RECOVERY AND PURIFICATION STRATEGIES

Due to the potential application and role in human nutrition and health, it is important that the recovery strategies for bioactive peptides do not compromise the biological activities. The research has shown that certain processing methods pursued on proteins affect the potency of bioactive peptides [12].

The biological materials are complex and there are differences in their properties that challenge the downstream bioprocess in the laboratory and the industry. This is due to the downstream purification and fractionation steps impact directly on the biological activities and therefore, on the economics of production [47]. The methodologies applied for the purification of proteins are generally applicable for peptide purification. However, it is important to consider that bonds breakage is specific depending on the source and method of extraction; hence, the obtained peptide fractions have their own peculiar properties. Also, is important to consider the application of the peptides, for example, Perez-Espitia [48] mentioned that some applications of peptides require low values of purity, between 70% and 95%. They also cited Ridge and Hettiarachchi [49] mentioning that for biological and structural studies as well as for therapeutic and clinical research, peptides must have 95% purity or greater.

In order to develop an efficient purification process, it is necessary to explore the effect of the isolation or purification method on the physical, chemical, and biological properties of bioactive peptides [12]. The peptide purification generally is carried out by a series of separation techniques. The separation processes are designed to isolate one or more individual components from a peptide mixture for further analytical identification, quantification, and application. Below are discussed some of the most used methods for the recovery and purification of peptides.

3.4.1 PROTEIN PRECIPITATION

The proteins obtained from the extraction process can be constituted by proteinaceous that range from high-molecular-weight proteins and also by peptide aggregates in the range of low-molecular-weight peptides [50]. The low-molecular-weight fraction represents a continuous size ranging from 1–10 kDa [51] to only a few amino acids [52].

The use of simple and rapid methodologies for the separation of proteins and peptides (high- and low-molecular-weight fractions) are needed. Selective precipitation is one of the most used techniques. The addition of acids such as trichloroacetic acid (TCA) [53], organic solvents [54], or the combination of precipitation agents [55], are often used for this purpose (Figure 3.1).

Ammonium sulfate, $(NH_4)_2SO_4$, is one of the most used precipitation agents due to its high solubility, which allows high ionic strength, low price, and availability in high purity [10]. The method consists of the removing of proteinaceous that easily aggregates from those that are very soluble making it a good initial purification step for small proteins and peptides. For example, the method has been used as a purification step for the characterization of antimicrobial peptides derived from the Japanese

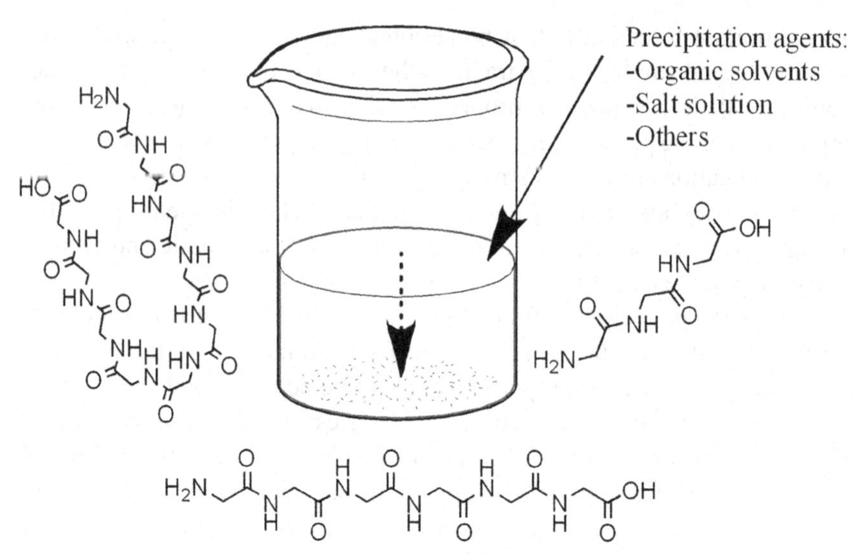

FIGURE 3.1 Schematic representation of peptides' precipitation process.

soybean fermented food (Natto) [56] and for the precipitation of the antimicrobial peptide cathelicidin-BF obtained from *Bacillus subtilis* [57].

The use of acetone is a common method for the precipitation and concentration of proteins. Simpson and Beynon [58] applied the precipitation of peptides by acetone and found that acetone, besides precipitation, modified peptides. The modification was selective predominantly on peptides which glycine residue was the second amino acid. Despite the protein loss is believed to be an inevitable consequence of acetone precipitation, some authors still using the method with modifications. For example, the use of ionic strength (1 to 100 mM NaCl) together with acetone (50 to 80%) maximizes the precipitation of proteins correlating the amount of salt with the amount of proteins in the sample [59].

The combination of techniques, such as acetone/TCA, are often used for the enhancing of peptides precipitation. Wu [60] developed a protocol for the precipitation of proteins from crops. They mentioned the main feature of the protocol was to maintain the advantages of both TCA/acetone precipitation, which removes non-protein compounds and selectively dissolves proteins, resulting in an effective protein purification step.

Selective precipitation of peptides is a real challenge for researchers and many strategies have been employed with that purpose. Rainer and Bonn [61] wrote an excellent review focused on the selective precipitation of phosphorylated peptides. They discussed the introduction in the last years of several non-chromatographic isolation technologies by using metal cations for the selective precipitation of phosphorylated peptides and proteins.

The isoelectric solubilization/precipitation technology is a useful tool for the obtaining of peptides and proteins from different sources. This technology was applied for the isolation of collagen hydrolysates from turkey heads and obtained by enzymatic hydrolysis. Depending on the used enzyme was the solubility of peptides. For example, low solubility was observed at pH 2 and 4 on peptides hydrolyzed with alcalase and trypsin, notwithstanding, was observed high solubility with low-molecular-weight peptides at pH 6 and 8 [44].

Another technique based on the precipitation of proteins and peptides is the isoelectric focusing (IEF). It is based on the separation of peptides/proteins solutions according to their isoelectric points (pI). The equipment has a focusing cell containing the mixture of peptides/proteins and a carrier ampholyte is submitted to an electric potential. It promotes the migration of the peptides/proteins to a position in an established pH gradient equivalent to their respective pI. The advantage of the method is that can fractionate a complex mixture of peptides according to the pI, but, has the disadvantage that overlapping between fractions because of their similar pI [62, 63].

3.4.2 ULTRAFILTRATION

As it was mentioned above, the purification of peptides generally requires the combination of purification techniques. The most used techniques for large-scale concentration and purification of peptides is the UF. This technology applies the principles of filtration, the separation of a mixture of components from a fluid by size differences, by using a membrane that acts as selective barrier controlling which component permeates and which are retained according to the molecular weight cut-off (Figure 3.2). The main characteristic of ultrafiltration is the application of hydraulic pressure to speed up the transport process [64].

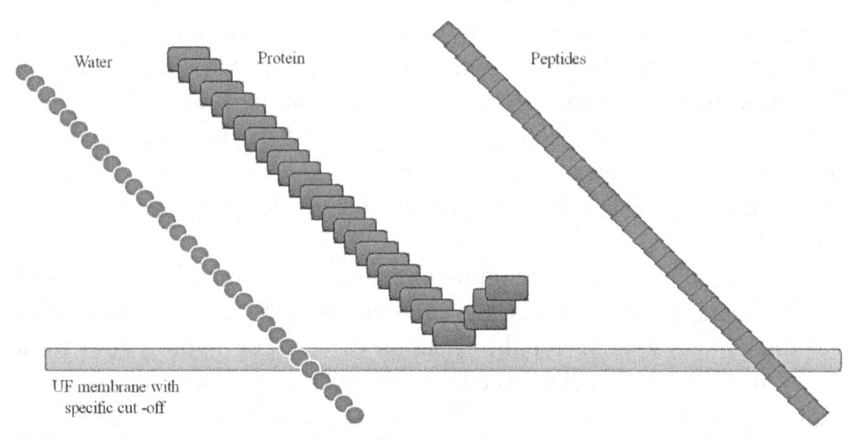

FIGURE 3.2 Representation of the UF process of peptides.

In the last years, the UF has been applied in the recovery of bioactive peptides obtained from by-products. For example, in the recovery of peptides from seafood by-products hydrolysates, the peptides were fractionated by molecular weight using UF membranes of different ranges of molecular weight (≥10, 5–10, 1–5 and ≤1 kDa) [65]. Besides, Abdelhedi [66] fractionated peptides obtained from smooth-hound viscera. They used membranes in the range of 50–1 kDa and characterized the peptides isolated. In both works mentioned previously, the peptides obtained were characterized *in silico* for the inhibition of angiotensin I-converting enzyme (ACE).

The UF has been applied also for the recovery of peptides from crops. Wang [67] isolated peptides from corn gluten meal by using membranes with molecular weight cut-off in the range of 50–6 kDa for the separation of proteins and peptides. The highest antioxidant activity was found in the fraction with molecular weight less than 6 kDa. They suggest that the antioxidant activity of peptides is related to the molecular weight and its absorption in the living body is relatively easy.

Despite being one of the most used methods for the separation of peptides and proteins, UF has the disadvantage that separates all the molecules in the cut-off range of membranes. Roblet [68] mentioned that ultrafiltration and nanofiltration are not selective for the large-scale separation of small peptides. They proposed the use of the electrodialysis with ultrafiltration membranes (EDUF) that allowed the separation of soybean peptides by molecular weight and global charges using electric fields as driving forces. The method is based on the separation of charged particles by their molecular charge and size in an electric field without any pressure. Then, peptides are selectively fractionated with no or minor collision of filtration membrane compared to the pressured filtration process. The EDUF is a trending method within the last decade and has been studied extensively in the separation and concentration of bioactive peptides from a variety of food proteins such as, β-lactoglobulin, soy, flaxseed, alfalfa white, and whey protein hydrolysates [9].

3.4.3 CHROMATOGRAPHY

Chromatographic methods are the most widely used for peptides purification. The clear example is the high-performance liquid chromatography (HPLC) and ultra-high-pressure liquid chromatography (UHPLC) [32]. HPLC and UHPLC have been applied for both purify and characterize peptides from different sources.

The HPLC in reversed phase mode (RP-HPLC) is used to separate peptides by hydrophobicity. On the other hand, the hydrophilic interaction is a useful method for the separation of hydrophilic substances and is based on the increasing of retention with increasing polarity of stationary phase and the decreasing of the polarity of the mobile phase is used for elution and is the opposite of RP-HPLC [63, 69]. This method was used for the separation of di-, tri-, and tetra-peptides [70].

Liu [2] developed a method for the identification of peptides derived from protein hydrolysates of *Mactra veneriformis.* The identification of 21 peptides was carried out by using a UHPLC-Q-TOF-MS equipment. The advantage of using UHPLC is the high throughput, high resolution and sensitivity [71] and generally has been used in tandem MS/MS for the identification of peptides.

The mass spectrometry is a very important tool in the process of identification of peptides. Different interfaces have been developed and they allowed the generation of ions from molecules with sensitivity to temperature. The electrospray ionization (ESI) and the matrix-assisted laser desorption-ionization time-of-flight (MALDI-TOF) are important ionization methods used for the identification and characterization of bioactive peptides and proteins by mass spectrometry [63].

Other wide used and efficient method for the purification of peptides is the size exclusion chromatography (SEC). The method is based on the fractionation of bioactive peptides by the retention of molecules on stationary phase according to their molecular size. The disadvantage of this method is the requirement of long columns or multiple columns in series for the separation of peptides mixtures. However, the method is considered soft because the impact on conformational structure is minimal [72].

Affinity chromatography and ion-exchange chromatography (IEC) are also wide used in purification of peptides. The first is based on the affinity of bioactive peptides to interact with specific stationary phase or solid support immobilized on a column. The interaction must be reversible in order to elute the peptides. The IEC is based on the interaction of charged peptides with packed supports bearing the opposite charges. The fractionation is carried out eluting cationic or anionic peptides by increasing charges, generally with salts. Desalting step is needed in order to eliminate salts [3].

3.5 CHARACTERIZATION

3.5.1 BIOACTIVITIES

The identification of new bioactive peptides is an important step towards the elucidation of new systems and technologies. Bioactive peptides adhere from their precursor proteins through limited excision since most of the occasions must undergo post-translational modifications to be biologically active [73]. Methods for the study of bioactive peptides are varied, depending on their amino acid sequence, they may have roles in different biological processes such as the evaluation of bioactivities such as antioxidative, antimicrobial [74], antimutagenic, antitumorigenic, antibacterial [75, 76].

3.5.1.1 ANTIOXIDATIVE ACTIVITIES

The uncontrolled production of free radicals that attack macromolecules such as membrane lipids, proteins, and DNA can lead to many health disorders [77]. Peptides with antioxidant activity can be obtained from the digestion of proteins of origin animal or vegetable, using endogenous or exogenous enzymes, microbial fermentation, processing and during gastrointestinal digestion [78]. Several mechanisms have been made for the antioxidant treatment properties of peptides, including metal ion chelation, free radical scavenging and aldehyde adduction [78, 79].

Peptides and food proteins have been found to possess free radical hardening properties involved in various states of oxidative insult [75]. Studies have reported different peptides with antioxidant activity with peptide sequence such as from the protein cow α_{s1}-casein (YFYPEL), cow β-casein (AVPYPQR), and cow β-lactoglobulin (MHIRL) all from cow milk [76], also Campos-Quevedo [80] reported peptides with antioxidant activity (Table 3.1). Peptides have shown efficacy against free radicals such as DPPH, superoxide, and hydroxyl radicals, ABTS radical scavenging [78, 79].

3.5.1.2 ANTIMICROBIAL PEPTIDES

The peptides with antimicrobial properties have been identified in distinct natural sources. Peptides with antimicrobial activity have shown inhibitory effects against microorganisms caused by food damage and invasion of a wide range such as bacteria, fungi, viruses, and parasites [75]. Bioactive peptides with antimicrobial activities in fermented foods, are produced by lactic acid bacteria (LAB). Peptides are good candidates as food additives [81].

3.5.2 TOXICOLOGY

Cell cultures assays have been performed which are useful for establishing dosage of peptides to exert beneficial antioxidant effects without cytotoxicity for in vivo experiments. Fish protein hydrolysates of flounder show antioxidant and cytoprotective activity against 2,2-azobis-(2-amidinopropane) dihydrochloride (AAPH) without cytotoxicity in a range of 12.5 to 200 μg/mL in Vero cells, monkey kidney fibroblast line; by showing effects such as cytotoxic protectors, and intracellular sweeping reactive oxygen species (ROS) [82].

Peptides originating from lactoferrin (LF) have been found to have a wide range of antimicrobial and anti-inflammatory activities in vivo, especially human LF f (1–11). This peptide has been studied for human safety and reported as safe (no adverse effects have been observed in markers or cytotoxicity) in healthy humans and in transplantation of hematopoietic stem cells in patients at doses as high as 5000 μg for a period of 5 days; however, there are still limitations because their pharmacokinetics is unknown [83].

Cyclotides are a family of plant peptides that are identified by having a macrocyclic backbone, in a study by He [84], showed that it has cytotoxic activity among other bioactivities. The cyclotides were isolated from *Viola philippica*, a plant of the Violaceae family; some of the cyclotides demonstrated cytotoxic activities against the MM96L, HeLa, and BGC–823 cancer cell lines [84].

The cytotoxic activity of oyster hydrolyzate (*Saccostrea cucullata*) was used in the study with the colon cancer cell line (HT–29) by the MTT assay which is based on the detection of mitochondrial dehydrogenase activity *in vivo*, under the same conditions, the hydrolyzate was evaluated in Vero (kidney epithelial cells of the African Green Monkey) with the aim of determining the effect on normal cells. Finding in the HT–29 cells an increased after 48 and 72 h of incubation the hydrolyzed oyster showed a remarkable dose and time-dependent cytotoxic effect on cancer cells where the inhibition of growth achieved up to 70% under conditions of 70 μg/ml of the sample [85]. Toxicity and bioavailability are evaluated primarily by using cell cultures, invertebrates or small animals using cells feasibility studies and determination of lethal dose 50 (LD50). The bioactive peptides derived from milk proteins are natural components that are normally considered safe for human consumption. Milk-specific bioactive peptides have been evaluated for cytotoxicity; in vitro cytotoxicity in rat, fibroblasts demonstrated low cytotoxicity over a period of 4–7 days of exposure, showing a $<30\%$ decrease in cell viability (Fibroblast L929) [86].

3.5.3 SEQUENCING

Bioactive peptides normally contain 3–20 amino acid residues per molecule [87]. For the sequencing of bioactive peptides, they have used equipment such as MALDI TOF and ESI (Peptide Sequencing by Mass Spectrometry) such as Samgina in 2009 where they studied the influence of different modifications of the N-terminal amino group in Phe3LeuPro2-NH2 natural peptide which was isolated from the skin secretion of the European tree frog *Hyla arborea* [88].

3.5.3.1 IDENTIFICATION OF PROTEINS BY PEPTIDE MAPPING

Through the strategy called mapping of peptide masses, a protein that is isolated is converted into peptides of size medium through the action of specific enzymes, mainly trypsin, but also V8, endolysin. Determination of the mass of the peptides of protein digestion allows comparisons to be made with the theoretical masses of the whole peptides expected for stored sequences in a database and suggest, those that fit more to the experimental data, Table 3.2 shows the tools for peptide mapping and sequencing [89].

TABLE 3.2 Some Tools That are Used in the Identification of Proteins by Means of Peptide Mapping or Sequencing of Peptides in Databases

Peptide mapping	
Mascot	http://www.matrixscience.com
Pepsa	www.pepsea.protana.com
Peptide sequencing	
InsPectT	http://peptide.ucsd.edu/
Mascot	http://www.matrixscience.com
ExPASy	http://www.expasy.org/
X!TANDEM	www.thegpm.org/TANDEM/

3.5.4 IN SILICO ANALYSIS AND MODELING

Several *in silico* peptide studies have been performed, Solanki [90], reported peptide sequences derived from the MASCOT program that compared to databases such as NCBI and identified it as protein (*Camelus dromedarius*), which comes from fermented camel milk, where purification of peptides was performed by RP-HPLC and identification of peptides by mass spectrometry (MS).

Recently, HPLC has been combined with mass spectrometry (MS) equipment, and liquid chromatography followed by tandem mass spectrometry (LC-MS/MS) this standard method for the characterization of the peptide sequences with which the structure of proteins and peptides is clarified, however, is a costly and long process method. Other important tools for the identification and characterization of proteins are electrospray ionization (ESI) and matrix-assisted laser desorption ionization (MALDI) [91]. Nonribosomal peptide synthesis (NRPS) has been used to produce peptides or other small molecules, furthermore, the mining of bacterial genomes through *in silico* analysis of NRPS genes offers an advantage for discovering novel ribosomal non-synthesized bioactive peptides [92].

The traditional protocol consists of performing sample purification, subsequently separating proteins or peptides by 2D-PAGE or RP-HPLC, an enzymatic digestion when proteins are intact and identifying by MS and MS/MS techniques, when are realized *in silico* it is carried out a protein digestion, and bioactive peptide database screening [93]. According to Giacometti, bioactive peptides are grouped into three large (> 25 AA), medium (7–25 AA), and small (<7 AA) peptides, most of which are found in small, most peptidic studies are performed by liquid chromatography, ionization-tandem mass spectrometry (LC-ESI-MS/MS) [34].

Bioinformatics is positioned to make an impact on bioactive peptide research, in the *in silico* approach the use of information deposited in databases, such as BIOPEP [94], PEPBANK or ERP-Moscow [95]. When the frequency of bioactive peptides is determined, the sequences can be obtained from the universal base which is (UniProtKB) [94]. Bioinformatics is a tool that can be combined with the classic for an exhaustive search of bioactive, peptide sequences or sustainable sources of its protein [94].

3.6 CONCLUDING REMARKS

The search of compounds with important biological activities is becoming trending to bioactive peptides. The wide sources and the relatively easy extraction have spread their use. New technologies have been used for the purification of peptides and the combination of two and/or more have allowed obtaining single and high purity peptides. The characterization of bioactive peptides by obtaining the aminoacidic sequences is important in order to understand the biological activities. In the last years, bioinformatics tools have been used for the characterization of *in silico* interaction of bioactive peptides with enzymes or proteins involved in specific diseases allowing to researchers to understand the processes and challenging to apply the knowledge to *in vivo* process.

KEYWORDS

- **bioactive peptides**
- **electrospray ionization**
- **high-performance liquid chromatography**
- **matrix-assisted laser desorption ionization**
- **nonribosomal peptide synthesis**
- **ultra-high-pressure liquid chromatography**

REFERENCES

1. Lafarga, T., & Hayes, M., (2014). Bioactive peptides from meat muscle and by-products: Generation, functionality, and application as functional ingredients. *Meat. Sci., 98* (2), 227–239. doi: 10.1016/j.meatsci.2014.05.036.
2. Liu, R., Zheng, W., Li, J., Wang, L., Wu, H., Wang, X., & Shi, L., (2015). Rapid identification of bioactive peptides with antioxidant activity from the enzymatic hydrolysate of *Mactra veneriformis* by UHPLC–Q-TOF mass spectrometry. *Food Chem., 167*, 484–489. doi: 10.1016/j.foodchem.2014.06.113.
3. Ortiz-Martinez, M., Winkler, R., & García-Lara, S., (2014). Preventive and therapeutic potential of peptides from cereals against cancer. *J. Proteomics., 111*, 165–183. doi: 10.1016/j.jprot.2014.03.044.
4. Sang, Y., & Blecha, F., (2008). Antimicrobial peptides and bacteriocins: Alternatives to traditional antibiotics. *Anim. Heal. Res. Rev.*, 9. doi: 10.1017/S1466252308001497.
5. Arroume, N., Froidevaux, R., Kapel, R., Cudennec, B., Ravallec, R., Flahaut, C. et al., (2016). Food peptides: Purification, identification and role in the metabolism. *Curr. Opin. Food Sci., 7*, 101–107. doi: 10.1016/j.cofs.2016.02.005.
6. Di Bernardini, R., Harnedy, P., Bolton, D., Kerry, J., O'Neill, E., Mullen, A. M. et al., (2011). Antioxidant and antimicrobial peptidic hydrolysates from muscle protein sources and by-products. *Food Chem., 124*(4), 1296–1307.
7. Ruan, G., Chen, Z., Wei, M., Liu, Y., Li, H., & Du, F., (2013). The study on microwave-assisted enzymatic digestion of ginkgo protein. *J. Mol. Catal. B Enzym., 94*, 23–28. doi: 10.1016/j.molcatb.2013.04.010.
8. Kadam, S. U., Tiwari, B. K., Álvarez, C., & O'Donnell, C. P., (2015). Ultrasound applications for the extraction, identification and delivery of food proteins and bioactive peptides. *Trends Food Sci. Technol., 46*(1), 60–67. doi: 10.1016/j.tifs.2015.07.012.

9. Suwal, S., Roblet, C., Amiot, J., Doyen, A., Beaulieu, L., Legault, J. et al., (2014). Recovery of valuable peptides from marine protein hydrolysate by electrodialysis with ultrafiltration membrane: Impact of ionic strength. *Food Res. Int.*, *65*, 407–415. doi: 10.1016/j.foodres.2014.06.031.
10. Duong-Ly, K. C., & Gabelli, S. B., (2014). Chapter seven-salting out of proteins using ammonium sulfate precipitation. *Methods Enzymol.*, *541*, 85–94.
11. Wu, S., Jia, S., Sun, D., Chen, M., Chen, X., Zhong, J. et al., (2005). Purification and characterization of two novel antimicrobial peptides subpeptin JM4-A and subpeptin JM4-B produced by *Bacillus subtilis* JM4. *Curr. Microbiol.*, *51*(5), 292–296. doi: 10.1007/s00284-005-0004-3.
12. Agyei, D., Ongkudon, C. M., Wei, C. Y., Chan, A. S., & Danquah, M. K., (2016). Bioprocess challenges to the isolation and purification of bioactive peptides. *Food Bioprod. Process*, *98*, 244–256. doi: 10.1016/j.fbp.2016.02.003.
13. Jiang, H., Tong, T., Sun, J., Xu, Y., Zhao, Z., & Liao, D., (2014). Purification and characterization of antioxidative peptides from round scad (*Decapterus maruadsi*) muscle protein hydrolysate. *Food Chem.*, *154*, 158–163. doi: 10.1016/j.foodchem.2013.12.074.
14. Capriotti, A. L., Caruso, G., Cavaliere, C., Samperi, R., Ventura, S., Zenezini Chiozzi, R. et al., (2015). Identification of potential bioactive peptides generated by simulated gastrointestinal digestion of soybean seeds and soy milk proteins. *J. Food Compos. Anal.*, *44*, 205–213. doi: 10.1016/j.jfca.2015.08.007.
15. Sakano, T., Notsumoto, S., Nagaoka, T., Morimoto, A., Fujimoto, K., Masuda, S. et al., (1988). Measurement of K vitamins in food by high-performance liquid chromatography with fluorometric detection. *Vitamins*, *62*.
16. Li-Chan, E. C., (2015). Bioactive peptides and protein hydrolysates: Research trends and challenges for application as nutraceuticals and functional food ingredients. *Curr. Opin. Food Sci.*, *1*, 28–37. doi: 10.1016/j.cofs.2014.09.005.
17. Park, S. Y., Lee, J. S., Baek, H. H., & Lee, H. G., (2010). Purification and characterization of antioxidant peptides from soy protein hydrolysate. *J. Food Biochem.*, *34*, 120–132. doi: 10.1111/j.1745–4514.2009.00313.x.
18. Das Neves, M. R. A., Campos, T., & Márquez, L. U. M., (2006). Modulação da pressão arterial por hidrolisados protéicos. *Brazilian J. Food Technol.*, *3*, 81–86.
19. Korhonen, H., (2009). Milk-derived bioactive peptides: From science to applications. *J. Funct. Foods*, *1*(2), 177–187.
20. Vercruysse, L., Van Camp, J., & Smagghe, G., (2005). ACE inhibitory peptides derived from enzymatic hydrolysates of animal muscle protein: A review. *J. Agric. Food Chem.*, *53*(21), 8106–8115.
21. Miguel, M., & Aleixandre, A., (2006). Antihypertensive peptides derived from egg proteins. *J. Nutr.*, *136*(6), 1457–1460.
22. Bougatef, A., Nedjar-Arroume, N., Ravallec-Plé, R., Leroy, Y., Guillochon, D., Barkia, A., ET AL., (2008). Angiotensin I-converting enzyme (ACE) inhibitory activities of sardinelle (*Sardinella aurita*) by-products protein hydrolysates obtained by treatment with microbial and visceral fish serine proteases. *Food Chem.*, *111*(2), 350–356. doi: 10.1016/j.foodchem.2008.03.074.

23. Yamamoto, K., (2001). Chemo-enzymatic synthesis of bioactive glycopeptide using microbial endoglycosidase. *J. Biosci. Bioeng.*, *92*(6), 493–501. doi: 10.1016/S1389–1723(01)80307–8.
24. Saitoh, E., Taniguchi, M., Ochiai, A., Kato, T., Imai, A., & Isemura, S., (2017). Bioactive peptides hidden in human salivary proteins. *J. Oral Biosci.*, *59*(2), 71–79. doi: 10.1016/j.job.2016.11.005.
25. Salas, C. E., Badillo-Corona, J. A., Ramírez-Sotelo, G., Oliver-Salvador, C., & Oliver-Salvador, C., (2015). Biologically active and antimicrobial peptides from plants. *Biomed Res. Int.*, 1–11. doi: 10.1155/2015/102129.
26. Wong, J. H., & Ng, T. B., (2005). Lunatusin, a trypsin-stable antimicrobial peptide from lima beans (*Phaseolus lunatus* L.). *Peptides*, *26*(11), 2086–2092. doi: 10.1016/j.peptides.2005.03.004.
27. Ambigaipalan, P., & Shahidi, F., (2017). Bioactive peptides from shrimp shell processing discards: Antioxidant and biological activities. *J. Funct. Foods*, *34*, 7–17. doi: 10.1016/j.jff.2017.04.013.
28. Abd El-Fattah, A. M., Sakr, S. S., El-Dieb, S. M., & Elkashef, H. A. S., (2016). Bioactive peptides with ACE-I and antioxidant activity produced from milk proteolysis. *Int. J. Food Prop.*, 1–10. doi: 10.1080/10942912.2016.1270963.
29. Nimalaratne, C., Bandara, N., & Wu, J., (2015). Purification and characterization of antioxidant peptides from enzymatically hydrolyzed chicken egg white. *Food Chem.*, *188*, 467–472. doi: 10.1016/j.foodchem.2015.05.014.
30. Mohanty, D., Jena, R., Choudhury, P. K., Pattnaik, R., Mohapatra, S., & Saini, M. R., (2016). Milk derived antimicrobial bioactive peptides: A review. *Int. J. Food Prop.*, *19*(4), 837–846 doi: 10.1080/10942912.2015.1048356.
31. Clemente, A., (2000). Enzymatic protein hydrolysates in human nutrition. *Trends Food Sci. Technol.*, *11*(7), 254–262. doi: 10.1016/S0924–2244(01)00007–3.
32. Singh, B. P., Vij, S., & Hati, S., (2014). Functional significance of bioactive peptides derived from soybean. *Peptides*, *54*, 171–179. doi: 10.1016/j.peptides.2014.01.022.
33. Azmir, J., Zaidul, I. S. M., Rahman, M. M., Sharif, K. M., Mohamed, A., Sahena, F. et al., (2013). Techniques for extraction of bioactive compounds from plant materials: A review. *J. Food Eng.*, *117*(4), 426–436. doi: 10.1016/j.jfoodeng.2013.01.014.
34. Giacometti, J., & Buretić-Tomljanović, A., (2017). Peptidomics as a tool for characterizing bioactive milk peptides. *Food Chem.*, *230*, 91–98. doi: 10.1016/j.foodchem.2017.03.016.
35. Heleno, S. A., Diz, P., Prieto, M. A., Barros, L., Rodrigues, A., Barreiro, M. F. et al., (2016). Optimization of ultrasound-assisted extraction to obtain mycosterols from *Agaricus bisporus* L. by response surface methodology and comparison with conventional soxhlet extraction. *Food Chem.*, *197*, 1054–1063. doi: 10.1016/j.foodchem.2015.11.108.
36. Niu, L., Li, J., Chen, M. S., & Xu, Z. F., (2014). Determination of oil contents in sacha inchi (Plukenetia volubilis) seeds at different developmental stages by two methods: Soxhlet extraction and time-domain nuclear magnetic resonance. *Ind. Crops Prod.*, *56*, 187–190. doi: 10.1016/j.indcrop.2014.03.007.
37. Da Porto, C., Porretto, E., & Decorti, D., (2013). Comparison of ultrasound-assisted extraction with conventional extraction methods of oil and polyphenols from grape

(*Vitis vinifera* L.) seeds. *Ultrason. Sonochem.*, *20*(4), 1076–1080. doi: 10.1016/j.ultsonch.2012.12.002.
38. Gros, C., Lanoisellé, J. L., & Vorobiev, E., (2003). Towards an alternative extraction process for linseed oil. *Chem. Eng. Res. Des.*, *81*(9), 1059–1065. doi: 10.1205/026387603770866182.
39. Zhao, R. J., Huo, C. Y., Qian, Y., Ren, D. F., & Lu, J., (2017). Ultra-high-pressure processing improves proteolysis and release of bioactive peptides with activation activities on alcohol metabolic enzymes *in vitro* from mushroom foot protein. *Food Chem.*, *231*, 25–32. doi: 10.1016/j.foodchem.2017.03.058.
40. Filly, A., Fernandez, X., Minuti, M., Visinoni, F., Cravotto, G., & Chemat, F., (2014). Solvent-free microwave extraction of essential oil from aromatic herbs: From laboratory to pilot and industrial scale. *Food Chem.*, *150*, 193–198. doi: 10.1016/j.foodchem.2013.10.139.
41. Puri, M., Sharma, D., & Barrow, C. J., (2012). Enzyme-assisted extraction of bioactives from plants. *Trends Biotechnol.*, *30*(1), 37–44. doi: 10.1016/j.tibtech.2011.06.014.
42. Liu, D., Chen, X., Huang, J., Huang, M., & Zhou, G., (2017). Generation of bioactive peptides from duck meat during post-mortem aging. *Food Chem.*, *237*, 408–415. doi: 10.1016/j.foodchem.2017.05.094.
43. Poojary, M. M., Orlien, V., Passamonti, P., & Olsen, K., (2017). Enzyme-assisted extraction enhancing the umami taste amino acids recovery from several cultivated mushrooms. *Food Chem.*, *234*, 236–244. doi: 10.1016/j.foodchem.2017.04.157.
44. Khiari, Z., Ndagijimana, M., & Betti, M., (2014). Low molecular weight bioactive peptides derived from the enzymatic hydrolysis of collagen after isoelectric solubilization/precipitation process of turkey by-products. *Poult. Sci.*, *93*(9), 2347–2362. doi: 10.3382/ps.2014–03953.
45. Nagendra, C. K. L., Manasa, D., Srinivas, P., & Sowbhagya, H. B., (2013). Enzyme-assisted extraction of bioactive compounds from ginger (*Zingiber officinale* Roscoe). *Food Chem.*, *139*(1–4), 509–514. doi: 10.1016/j.foodchem.2013.01.099.
46. Campbell, K. A., Vaca-Medina, G., Glatz, C. E., & Pontalier, P. Y., (2016). Parameters affecting the enzyme-assisted aqueous extraction of extruded sunflower meal. *Food Chem.*, *208*, 245–251. doi: 10.1016/j.foodchem.2016.03.098.
47. Agyei, D., Potumarthi, R., & Danquah, M. K., (2013). Production of lactobacilli proteinases for the manufacture of bioactive peptides: Part II-downstream processes. *Mar. Proteins Pept. Biol. Act. Appl.*, 231–251.
48. Perez, E. P. J., De Fatima, F. S. N., Dos Reis, C. J. S., De Andrade, N. J., Souza, C. R., & Alves, M. E. A., (2012). Bioactive peptides: Synthesis, properties, and applications in the packaging and preservation of food. *Compr. Rev. Food Sci. Food Saf.*, *11*(2), 187–204. doi: 10.1111/j.1541–4337.2011.00179.x.
49. Ridge, S., & Hettiarachchi, K., (1998). Peptide purity and counter ion determination of bradykinin by high-performance liquid chromatography and capillary electrophoresis. *J. Chromatogr. A.*, *817*(1), 215–222.
50. Dallas, D. C., Guerrero, A., Parker, E. A., Robinson, R. C., Gan, J., German, J. B. et al., (2015). Current peptidomics: Applications, purification, identification, quantification, and functional analysis. *Proteomics.*, *15*(5–6), 1026–1038.

51. Hortin, G. L., Shen, R. F., Martin, B. M., & Remaley, A. T., (2006). Diverse range of small peptides associated with high-density lipoprotein. *Biochem. Biophys. Res. Commun.*, *340*(3), 909–915.
52. Lahrichi, S. L., Affolter, M., Zolezzi, I. S., & Panchaud, A., (2013). Food peptidomics: Large scale analysis of small bioactive peptides-a pilot study. *J. Proteomics.*, *88*, 83–91.
53. Dallas, D. C., Guerrero, A., Khaldi, N., Castillo, P. A., Martin, W. F., Smilowitz, J. T. et al., (2013). Extensive *in vivo* human milk peptidomics reveals specific proteolysis yielding protective antimicrobial peptides. *J. Proteome Res.*, *12*(5), 2295–2304.
54. Manes, N. P., Gustin, J. K., Rue, J., Mottaz, H. M., Purvine, S. O., Norbeck, A. D. et al., (2007). Targeted protein degradation by salmonella under phagosome-mimicking culture conditions investigated using comparative peptidomics. *Mol. Cell. Proteomics*, *6*(4), 717–727.
55. Khmelnitsky, Y. L., Belova, A. B., Levashov, A. V, & Mozhaev, V. V., (1991). Relationship between Surface hydrophilicity of a protein and its stability against denaturation by organic solvents. *FEBS Lett.*, *284*(2), 267–269.
56. Kitagawa, M., Shiraishi, T., Yamamoto, S., Kutomi, R., Ohkoshi, Y., Sato, T. et al., (2017). Novel antimicrobial activities of a peptide derived from a Japanese soybean fermented food, natto, against streptococcus pneumoniae and bacillus subtilis group strains. *AMB Express.*, *7*(1), 127. doi: 10.1186/s13568-017-0430-1.
57. Luan, C., Zhang, H. W., Song, D. G., Xie, Y. G., Feng, J., & Wang, Y. Z., (2014). Expressing antimicrobial peptide cathelicidin-BF in *Bacillus subtilis* using SUMO technology. *Appl. Microbiol. Biotechnol.*, *98*(8), 3651–3658. doi: 10.1007/s00253-013-5246-6.
58. Simpson, D. M., & Beynon, R. J., (2010). Acetone precipitation of proteins and the modification of peptides. *J. Proteome Res.*, *9*(1), 444–450. doi: 10.1021/pr900806x.
59. Crowell, A. M. J., Wall, M. J., & Doucette, A. A., (2013). Maximizing recovery of water-soluble proteins through acetone precipitation. *Anal. Chim. Acta*, *796*, 48–54.
60. Wu, X., Xiong, E., Wang, W., Scali, M., & Cresti, M., (2014). Universal sample preparation method integrating trichloroacetic acid/acetone precipitation with phenol extraction for crop proteomic analysis. *Nat. Protoc.*, *9*(2), 362.
61. Rainer, M., & Bonn, G. K., (2015). Enrichment of phosphorylated peptides and proteins by selective precipitation methods. *Bioanalysis*, *7*(2), 243–252. doi: 10.4155/bio.14.281.
62. Guijarro-Díez, M., García, M. C., Crego, A. L., & Marina, M. L., (2014). Off-line two dimensional isoelectrofocusing-liquid chromatography/mass spectrometry (Time of flight) for the determination of the bioactive peptide lunasin. *J. Chromatogr. A.*, *1371*, 117–124.
63. Janser, R., Castro, S. De, & Sato, H. H., (2015). Biologically active peptides : Processes for their generation, Puri Fi cation and Identi Fi cation and applications as natural additives in the food and pharmaceutical industries. *Frin.*, *74*, 185–198. doi: 10.1016/j.foodres.2015.05.013.
64. Cheryan, M., (1998). *Ultrafiltration and Microfiltration Handbook* (Vol. 1; p, 6). Technomic Pub. Co: Boca Raton, Florida.

65. Ngo, D. H., Vo, T. S., Ryu, B., & Kim, S. K., (2016). Angiotensin- I- converting enzyme (ACE) inhibitory peptides from pacific cod skin gelatin using ultrafiltration membranes. *Process Biochem.*, *51*(10), 1622–1628. doi: 10.1016/j.procbio.2016.07.006.
66. Abdelhedi, O., Nasri, R., Mora, L., Jridi, M., Toldrá, F., & Nasri, M., (2018). In silico analysis and molecular docking study of angiotensin I-converting enzyme inhibitory peptides from smooth-hound viscera protein hydrolysates fractionated by ultrafiltration. *Food Chem.*, *239*, 453–463. doi: 10.1016/j.foodchem.2017.06.112.
67. Wang, X., Zheng, X., Kopparapu, N., Cong, W., Deng, Y., Sun, X. et al., (2014). Purification and evaluation of a novel antioxidant peptide from corn protein hydrolysate. *Process Biochem.*, *4*(9), 1562–1569. doi: 10.1016/j.procbio.2014.05.014.
68. Roblet, C., Doyen, A., Amiot, J., Pilon, G., Marette, A., & Bazinet, L., (2014). Enhancement of glucose uptake in muscular cell by soybean charged peptides isolated by electrodialysis with ultrafiltration membranes (EDUF). Activation of the AMPK pathway. *Food Chem.*, *147*, 124–130.
69. Yoshida, T., (2004). Peptide separation by hydrophilic-interaction chromatography: A review. *J. Biochem. Biophys. Methods*, *60*(3), 265–280. doi: 10.1016/j.jbbm.2004.01.006.
70. Le Maux, S., Nongonierma, A. B., & FitzGerald, R. J., (2015). Improved short peptide identification using HILIC–MS/MS: Retention time prediction model based on the impact of amino acid position in the peptide sequence. *Food Chem.*, *173*, 847–854. doi: 10.1016/j.foodchem.2014.10.104.
71. Fekete, S., & Guillarme, D., (2014). Ultra-high-performance liquid chromatography for the characterization of therapeutic proteins. *TrAC. Trends Anal. Chem.*, *63*, 76–84. doi: 10.1016/j.trac.2014.05.012.
72. Mora, L., Escudero, E., Fraser, P. D., Aristoy, M. C., & Toldrá, F., (2014). Proteomic identification of antioxidant peptides from 400 to 2500Da generated in spanish dry-cured ham contained in a size-exclusion chromatography fraction. *Food Res. Int.*, *56*, 68–76. doi: 10.1016/j.foodres.2013.12.001.
73. Toshinai, K., Nakazato, M., Medicine, I., & Medicine, F., (2009). Review neuroendocrine regulatory peptide-1 and -2 : Novel bioactive peptides processed from VGF., *66*, 1939–1945. doi: 10.1007/s00018-009-8796-0.
74. Capriotti, A. L., Cavaliere, C., Foglia, P., Piovesana, S., Samperi, R., Zenezini Chiozzi, R. et al., (2015). Development of an analytical strategy for the identification of potential bioactive peptides generated by *in vitro* tryptic digestion of fish muscle proteins. *Anal. Bioanal. Chem.*, *407*(3), 845–854. doi: 10.1007/s00216-014-8094-z.
75. Kannan, A., Hettiarachchy, N., & Marshall, M., (2012). Food proteins and peptides as bioactive agents. *Bioactive Food Proteins and Peptides*, (Vol. 1; pp. 1–28). CRC Press: Boca Raton, Florida.
76. Capriotti, A. L., Cavaliere, C., Piovesana, S., Samperi, R., & Laganá, A., (2016). Recent trends in the analysis of bioactive peptides in milk and dairy products. *Anal. Bioanal. Chem.*, *408*(11), 2677–2685. doi: 10.1007/s00216–016–9303–8.
77. Kim, S. K., & Wijesekara, I., (2010). Development and biological activities of marine-derived bioactive peptides: A review. *J. Funct. Foods*, *2*(1), 1–9. doi: 10.1016/j.jff.2010.01.003.

78. Samaranayaka, A. G. P., & Li-Chan, E. C. Y., (2011). Food-derived peptidic antioxidants: A review of their production, assessment, and potential applications. *J. Funct. Foods*, *3*(4), 229–254. doi: 10.1016/j.jff.2011.05.006.
79. Irshad, I., Kanekanian, A., Peters, A., & Masud, T., (2015). Antioxidant activity of bioactive peptides derived from bovine casein hydrolysate fractions. *J. Food Sci. Technol.*, *52*(1), 231–239. doi: 10.1007/s13197–012–0920–8.
80. Campos-Quevedo, N., Rosales-Mendoza, Sergio Paz-Maldonado, L. M., Martinez-Salgado, Luis, Guevara-Arauza, & Carlos, S. G. R. E., (2013). Production of milk-derived bioactive peptides as precursor chimeric proteins in chloroplasts of *Chlamydomonas reinhardtii*. *Plant Cell. Tissue Organ Cult.*, *113*, 217–225. doi: 10.1007/s11240–012–0261–3.
81. Gálvez, A., Abriouel, H., López, R. L., & Omar, N. Ben., (2007). Bacteriocin-based strategies for food biopreservation. *Int. J. Food Microbiol.*, *120*(1–2), 51–70. doi: 10.1016/j.ijfoodmicro.2007.06.001.
82. Chakrabarti, S., Forough, J., Wu, J., Jahandideh, F., & Wu, J., (2014). Food-derived bioactive peptides on inflammation and oxidative stress. *Biomed Res. Int.*, doi: 10.1155/2014/608979.
83. Nongonierma, A. B., & Fitz Gerald, R. J., (2015). The scientific evidence for the role of milk protein-derived bioactive peptides in humans: A review. *J. Funct. Foods*, *17*, 640–656. doi: 10.1016/j.jff.2015.06.021.
84. He, W., Chan, L. Y., Zeng, G., Daly, N. L., Craik, D. J., & Tan, N., (2011). Isolation and characterization of cytotoxic cyclotides from viola philippica. *Peptides*, *32*(8), 1719–1723. doi: 10.1016/j.peptides.2011.06.016.
85. Umayaparvathi, S., Arumugam, M., Meenakshi, S., Dräger, G., Kirschning, A., & Balasubramanian, T., (2014). Purification and characterization of antioxidant peptides from oyster (*Saccostrea cucullata*) hydrolysate and the anticancer activity of hydrolysate on human colon cancer cell lines. *Int. J. Pept. Res. Ther.*, *20*(2), 231–243. doi: 10.1007/s10989–013–9385–5.
86. Nongonierma, A. B., & FitzGerald, R. J., (2016). Strategies for the discovery, identification and validation of milk protein-derived bioactive peptides. *Trends Food Sci. Technol.*, *50*, 26–43. doi: 10.1016/j.tifs.2016.01.022.
87. Hati, S., Sakure, A., & Mandal, S., (2016). Impact of proteolytic lactobacillus helveticus MTCC5463 on production of bioactive peptides derived from honey based fermented milk. *Int. J. Pept. Res. Ther.*, 1–7. doi: 10.1007/s10989–016–9561–5.
88. Samgina, T. Y., Kovalev, S. V., Gorshkov, V. A., Artemenko, K. A., Poljakov, N. B., & Lebedev, A. T., (2010). N-terminal tagging strategy for de novo sequencing of short peptides by ESI-MS/MS and MALDI-MS/MS. *J. Am. Soc. Mass Spectrom.*, *21*(1), 104–111. doi: 10.1016/j.jasms.2009.09.008.
89. Abián, J., Carrascal, M., & Gay, M., (2008). Universidad complutense de madrid departamento de bioquímica y biología molecular tesis doctoral fernando de la cuesta marina directores. *Proteómica, 2*, 17–31.
90. Solanki, D., Hati, S., & Sakure, A., (2017). *In silico* and *in vitro* analysis of novel angiotensin i-converting enzyme (ACE) inhibitory bioactive peptides derived from fermented camel milk (*Camelus dromedarius*). *Int. J. Pept. Res. Ther.*, *23* (4), 441–459. doi: 10.1007/s10989–017–9577–5.

91. Wang, X., Yu, H., Xing, R., & Li, P., (2017). Review article characterization, preparation, and purification of marine bioactive peptides. *Biomed Res. Int.,* 1–16. doi:10.1155/2017/9746720.
92. Royer, M., Koebnik, R., Marguerettaz, M., Barbe, V., Robin, G. P., Brin, C. et al., (2013). Genome mining reveals the genus *xanthomonas* to be a promising reservoir for new bioactive non-ribosomally synthesized peptides. *BMC Genomics*, *14*, 658. doi: 10.1186/1471–2164–14–658.
93. Saavedra, L., Hebert, E. M., Minahk, C., & Ferranti, P., (2013). An overview of "omic" analytical methods applied in bioactive peptide studies. *Food Res. Int.*, *54*(1), 925–934. doi: 10.1016/j.foodres.2013.02.034.
94. Udenigwe, C. C., (2014). Bioinformatics approaches, prospects and challenges of food bioactive peptide research. *Trends Food Sci. Technol.*, *36*(2), 137–143. doi: 10.1016/j.tifs.2014.02.004.
95. Mora, L., Gallego, M., Reig, M., & Toldrá, F., (2017). Challenges in the quantitation of naturally generated bioactive peptides in processed meats. *Trends Food Sci. Technol., 69*, 306–314. doi: 10.1016/j.tifs.2017.04.011.

CHAPTER 4

AGAVE FRUCTANS AS FUNCTIONAL COMPONENTS IN FOODS: IMPORTANCE, APPLICATIONS, AND DETERMINATION METHODS

SARAI ESCOBEDO-GARCÍA, ADRIANA C. FLORES-GALLEGOS, JUAN C. CONTRERAS-ESQUIVEL, JESÚS A. SALAS-TOVAR, CRISTÓBAL N. AGUILAR, and RAÚL RODRÍGUEZ-HERRERA

Food Research Department, School of Chemistry, Autonomous University of Coahuila, Boulevard Venustiano Carranza and José Cárdenas s/n, República Oriente, Saltillo 25280, Coahuila, México, E-mail: raul.rodriguez@uadec.edu.mx

ABSTRACT

Agaves are plants mainly distributed in America; these plants have the capacity for adaptation to grow at different environmental conditions. In Mexico, Agaves have a high economic importance because of the many products that can be obtained from them. Agaves contain carbohydrates that have been used to obtain alcoholic beverages. Fructans are the principal carbohydrates found in agaves. The composition and structure of fructans from *Agave* species have been widely described in many studies. Agave fructans have demonstrated to possess a prebiotic effect. Several techniques have been used for fructans analysis. Chromatographic methods are the most commonly used for fructooligosaccharides and fructans analyses because other methods are complicated and expensive, and because of lack of standards. This paper describes the importance of fructans as food functional component, applications, and compares different chromatographic and other techniques for agave fructans analysis.

4.1 INTRODUCTION

Mexico is considered the center of origin and diversity of the *Agavaceae* family. This family consists of eight genera and at least 295 species of this family have been described [1]. Since ancient times, plants of *Agave* genus have a wide variety of applications; they have been used mainly for the production of traditional alcoholic beverages such as tequila and mescal because of its high reserve carbohydrate content [2, 3]. Since the past century, many studies have been made to determinate fructans content in different plant species. The potential market for agave fructans is estimated as the same as inulin from other sources as well as that of soluble fibers (around 100,000 tons of inulin and 200,000 tons of fructooligosaccharides worldwide in 2009) [4]. Fructans have been found in Angiosperms, in several families of both monocots and dicots such as *Liliaceae*, *Amaryllidaceae*, *Gramineae*, *Compositae*, *Nolinaceae* including *Agavaceae* [5, 6]. Fructans are polymers of fructose linked by fructose-fructose glycosidic bonds, these are known as fructooligosaccharides (FOS) which are composed of 2 to 10 fructose molecules, and nonetheless an appropriate way to identify them is that they are polysaccharides with a degree of polymerization (DP) with more of 10 molecules of fructose in the chain [7]. These are the principal products formed during photosynthetic Crassulacean acid metabolism (CAM) for CO_2 fixation, which is a biochemical and physiological plants adaptation to survive and adapt to arid and semiarid regions where they grow. During photosynthesis, glucose is synthesized in the plant by the action of a phosphatase on glucose–1-phosphate. Thereafter, glucose is transformed into glucose–6-phosphate by a hexokinase; later this product is converted into fructose-6-phosphate by hexose phosphate isomerase. Finally, this last product loses the phosphate group and releases fructose. Glucose and fructose are the most important energy supply by plants [3, 8].

Several plants use fructans as a reserve like starch or sucrose, and store both fructans and starch. More than 60% of carbohydrates in dry basis present in those plants are fructans, which are mainly stored in leaves, stems, and roots. In grasses, they are principally stored in the leaf base and used for re-growth after defoliation, while starch is mostly stored in seeds. Fructans are synthesized and stored in vacuole whereas starch is stored in specialized organelles (plastids) of plant cells [9–11]. After starch, fructans are the most abundant nonstructural polysaccharides found in 15%

of higher plants as well as in a wide variety of bacteria and fungi [12] in plants, fructans are used as reserve carbohydrates [13]. In dicots, inulin is the only fructan type used to cover energy demand, and is stored in reserve vegetative organs, whereas in monocots, a cryoprotective role of fructans has been demonstrated in oat, wheat, and other cereals [14, 15]. In addition, fructans have been involved in plant stress tolerance such as drought and freezing tolerance and also they play a fundamental role in osmoregulation [16]. González-Cruz et al. [17] established a relationship between plant age and fructan synthesis in *A. atrovirens* Karw, they identified changes in fructan composition in leaves at three ages (3, 6 and 9 years) and also, they quantified the fructosyltransferase enzymatic complex activity, thus determining the growth stage with the highest fructan content. The highest content of non-structural carbohydrates was in the leaves of youngest plants and decreased by 86% in those of the oldest ones. In the youngest plants, the main non-structural carbohydrate was sucrose, meanwhile in the oldest plants predominated inulin-type fructans, glucose, and fructose [18].

Fructans are classified according to the fructosyl bond into inulins, levans, graminans, and fructan neoseries. Inulins contain β (2→1), levans β (2→6), graminans have both types bonds and fructan neoseries which are featured by an internal glucose molecule that can be lengthened by β (2→1) and/or (2→6) bonds, generating inulin and/or levan neoseries (Figure 4.1), respectively [1, 16].

López et al. [8] showed that agave fructans are complex mixtures of fructans containing β (2→ 1) and β (2→ 6) linkages and external (graminans) and internal glucose (neoseries fructan) units and are found in *Agave tequilana*, *A. angustifolia*, *A. potatorum*, *A. salmiana* and *A. fourcroydes*; these fructans were called agavins [19].

4.2 APPLICATIONS

Because of β-configuration in their fructose monomers, the human digestive enzymes (α-glucosidase, maltase, isomaltase, and sucrase) cannot hydrolyze fructans, therefore these are classified as non-digestible oligosaccharides. Because of this reason, a fructan-rich diet may have health-promoting effects, also these compounds are considered as prebiotic [20]. A prebiotic is a food ingredient that beneficially affects the host by

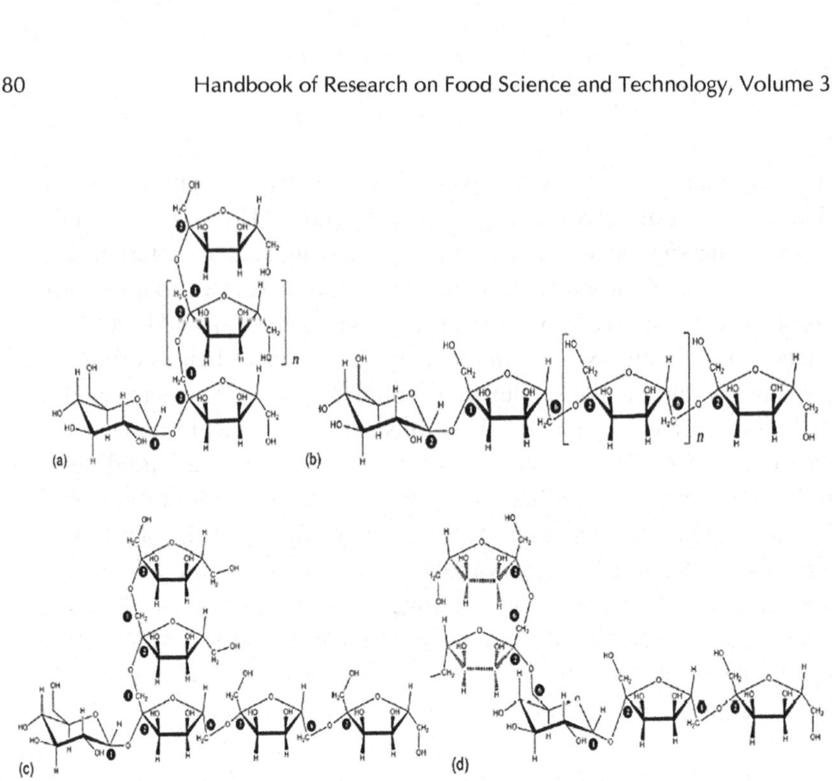

FIGURE 4.1 Fructan structures: (a) inulin, (b) levan, (c) graminan, (d) neoseries (adapted from Ref. [75]).

selectively stimulating growth and/or activity of one or a limited number of bacteria in the colon, which improves host health. Recently, it was found that many oligosaccharides and polysaccharides (including dietary fiber) in food have prebiotic effect; nevertheless, not all dietary carbohydrates are prebiotics. To consider a carbohydrate as prebiotic, this have to resist gastric acidity, hydrolysis by gastric enzymes and gastrointestinal absorption, be fermented by the intestinal microbiota and promote selectively growth of intestinal microbiota associated with host health [21, 22].

There are different reports mainly with lactobacilli and bifidobacteria strains, which evaluated the prebiotic effect of agave fructans. Urías-Silvas and López [20] mentioned the potential as prebiotics of five *Agave* species (*A. tequilana*, *A. angustifolia*, *A. potatorum*, *A. cantala* and *A. fourcroydes*), *Dasylirion* sp. and a commercial inulin-type fructans were tested as substrate for lactobacilli and bifidobacteria strains growth and it was found that these kinds of microorganisms can grow using *Agave* and *Dasylirion* fructans as carbon source. Most fructans stimulate the growth of bacteria strains more effectively than commercial inulin. In addition, Santiago-García and López [23] studied the prebiotic effect of fructans and mixtures with different DP of *A. angustifolia* using bifidobacteria and lactobacilli strains. They reported the prebiotic potential and besides that short-DP fructans in mixtures influenced rate of fermentation by probiotic bacteria. In another work, using bifidobacteria and lactobacilli strains, Velázquez-Martínez et al. [24] found similar results using *A. angustifolia* fructan fractions as carbon source for probiotic bacteria and the most assimilated fraction was the fraction with low DP.

Fructans effect on health has been tested with animal's lab. Urías-Silvas et al. [25] described the physiological effects of agavins from *A. tequilana* which demonstrated significant effects on glucose and lipid homeostasis in male mice C57BL/6J. Santiago-García and López [26] evaluated the effect of a supplemented diet with 10% agavins which has short DP from *A. angustifolia* Haw. and *A. potatorum* Zucc. along with chicory fructans. Mice fed with a fructan diet showed lower energy intake, body weight gain and triglycerides than mice fed with a standard diet. They attributed capability of agavins to enhance glucose and lipid homeostasis by production of SCFA's (short-chain fatty acids) in the gut and by secretion of peptides implicated in appetite regulation. These results showed that agavins reduce food intake and might help to control obesity and metabolic disorders.

Other research reveals that fructans might help to mitigate bone loss and improve bone formation. García-Vieyra et al. [27] indicated that agave fructans have an effect on mineral absorption improvement. They used C57BL/6J mice and fed them with standards diets, diets supplemented with 10% agave fructans or 10% inulin fructans. Calcium, magnesium, and osteocalcin levels were evaluated. Levels of calcium in plasma and bone increased in mice fed with agave fructans, and also osteocalcin increased. Scanning electron microscopy showed that agave fructans help to mitigate bone loss in mice. Nevertheless, mechanism of increased mineral absorption by SCFA's is not completely understood yet.

García-Curbelo et al. [28] evaluated the prebiotic effect of *A. fourcroydes* fructans on male mice (C57BL/6J) that were fed with standard diet or diet supplemented with 10% of *Cichorium intybus* fructans (Raftilose P95) and agave fructans. They monitored the body weight, food intake, blood glucose, triglycerides, and cholesterol, gastrointestinal organ weights, fermentation indicators in cecal and colon contents and mineral contents in femur. Results suggested that diet supplemented with agave fructans produced a prebiotic response similar to or greater than Raftilose P95, and additionally, a reduction of serum glucose, triglycerides, cholesterol and increased mineral concentrations in femurs. Also, Huazano-García and López [29] fed male mice (C57BL/6J) with a standard diet or high-fat diet over 5 weeks to induce over-weightiness in mice, then they fed mice with a diet shift and a diet shift supplemented with agavins or inulin for additional 5 weeks. Solely, the overweight mice fed with agavins or inulin showed a reverse in metabolic disorders caused by a high-fat diet. Moreover, consumption of agavins or inulin increased SCFA concentrations and modulated hormones implicated in food intake regulation. Another work with mice showed that a diet supplemented with 10% *A. tequilana* agavins decreased body weight gain which is induced by a high-fat diet, increased production of SCFA and also might act as bioactive ingredients with anti-oxidant and protective roles in the brain [30]. Also, there are studies in humans, Holscher et al. [31] evaluated the gastrointestinal tolerance using 5 and 7.5 g per day of agave inulin in healthy adults. Results indicated that doses up to 7.5 g per day of *Agave* inulin reduce gastrointestinal upset, did not increase diarrhea and enhance laxation in healthy adults. In addition, other studies have been demonstrated the prebiotic capacity of Agave fructans. López-Velázquez et al. [32] worked with newborns to demonstrate the efficacy of agave fructans from *A. tequilana* Weber as prebiotics. They found that agave fructans promote a bifidogenic effect and regulate triglycerides, cholesterol, and lipoproteins.

Zamora-Gasga et al. [33] used *A. tequilana* by-products to prepare oat-based granola bars, results showed that these by-products might be used as functional ingredients and also could maintain consumer preference. Zamora-Gasga et al. [34] evaluated an *in vitro* enzymatic digestion process of a granola bar with agave fructans as ingredients. They found that addition of agave fructans increased SCFA production and metabolite production. Agave fructans may improve food sensorial characteristics. Crispín-Isidro et al. [35] tested the microstructural, rheological, and sensory properties of reduced-fat stirred yogurts added with inulin and agave fructans in comparison with those of a full-fat control yogurt. Results indicated that yogurts added with inulin and agave fructans showed sensory characteristics (viscosity, creaminess, flavor, and overall acceptability) higher than those full-fat control yogurts. In summary, Agave fructans called agavins function as non-digestible dietary fiber, as prebiotic, they are able to promote growth of bifidobacteria and lactobacilli as well as production of short chain fatty acids (SCFA) which are considered indicators of healthy intestinal microbiota, also to mitigate bone loss and other applications.

4.3 AGAVE FRUCTANS EXTRACTION

Because to the multiple applications of agave fructans in the food industry, and the wide distribution in nature; extraction, purification, and characterization of these compounds, is increasing in recent years. Many investigations have been performed to set optimum extraction conditions of inulin-type fructans from plants. The most important factors that have an influence on yield are temperature, extraction time and solvent/solid ratio. The solubility of inulin *significantly* increases in hot water, being almost insoluble at 25°C and more soluble at 90°C, hence, the industrial production process is based on diffusion in hot water [36].

González-Cruz et al. [37] described three different non-structural carbohydrate extraction methods. They used *A. atrovirens* karw pines to extract inulin-type fructans, pines were grounded in an industrial blender and juice was filtered through a cotton cloth producing raw extract; the second method was carried out placing pines in water boiled for 20 min, and then they were grounded and obtained the aqueous extract, [38] the last extraction was obtained placing pines in 70% ethanol for 20 min, after that, they were grounded and obtained the alcohol extracts [19]. Results showed that the best extraction method was obtained with aqueous extract, but that both aqueous and alcohol extracts were inadequate for extraction of 1-kestose. Dalonso et al. [39] demonstrated that long chain inulin can be precipitated from aqueous solutions in presence of high concentrations of organic solvents such as methanol, ethanol, propanol, acetonitrile, and acetone among others. The best organic solvent for maintenance of natural DP was acetone followed by ethanol and methanol. Acetone possesses ability to remove water from solvation of polysaccharides, thereby enabling dehydration and subsequent precipitation. Table 4.1 shows the most common methods for agave fructans extraction.

Moreover, ultrasound-assisted extraction has been recently proposed to improve inulin-type fructans extraction in comparison with traditional methods, nevertheless, caution is needed in using sonication to extract inulin-type fructans, because some low-molecular-weight fragments

TABLE 4.1 Agave Fructans Extraction Methods

Plant	Plant's treatment	Extraction	Reference
A. tequilana Weber var. azul	Agave pine was milled and the pulp solid fraction was stored at –20°C.	Fructans of agave pines were extracted with 80% ethanol shaking for 75°C/1 h. Then filtered and re-extracted with water at 70°C/30 min.	López et al. [8]
A. tequilana Weber var. azul, *A. angustifolia*, *A. cantala* and *A. fourcroydes*	-	Fructans were extracted from stems, stems were extracted twice with 80% ethanol shaking 55°C/1 h. samples were filtered and re-extracted with water at 55°C/30 min.	Urías-Silvas and López [20]
A. americana	Tissues were homogenized with distilled water.	Fructans were extracted from by heating the homogenate and shaking at 75°C/30 min.	Ravenscroft et al. [57]
A. tequilana Weber var. azul	Agave heads were frozen at –20°C.	Fructans were extracted mixing the samples with distilled water. The mix was blended and stirred at 70°C/7 h.	Arrizon et al. [58]
A. tequilana Weber var. azul	-	The fructans were extracted with water at 70°C/2 h.	Ávila-Fernández et al. [72]
A. atrovirens Karw	Leaves and pines were frozen	*Leaves/pines were grounded and blended, and then the juice was filtered. *Leaves/pines were placed in water boiled for 20 min and then grounded and filtered. *Leaves/pines were placed in 70% ethanol for 20 min and then grounded and filtered.	González-Cruz et al. [17, 37]
A. tequilana Weber var. azul	-	They extracted the fructans with water at 80°C.	Espinosa-Andrews and Urías-Silvas [73]
A. tequilana Weber var. azul	Agave heads were frozen and thawed, after that, juice was extracted and finally centrifuged 5000 rpm/15 min.	Fructans were extracted precipitating with ethanol (20, 40, 60 and 80%). The obtained precipitate was pelleted for 15 min and then decanted.	Montañez-Soto et al. [74]

TABLE 4.1 *(Continued)*

Plant	Plant's treatment	Extraction	Reference
A. angustifolia and *A. potatorum*	-	Stems were extracted twice using 80% ethanol at 55°C/1 h. Sample was filtered and plant material re-extracted with water 55°C/60 min.	Santiago-García and López [26]
A. angustifolia	-	Samples were stirred continuously in water at 80°C. Then the aqueous extract was microfiltered and ultra-filtered using 10 KDa MWCO hollow fiber cartridge (Amicon™, Merck Millipore, Darmstadt, Germany).	Velázquez-Martínez et al. [24]
A. tequilana Weber var. azul	-	Fructan extraction was carried out in water at 80°C shaking for 15 min/150 rpm, after that sample was filtered.	González-Herrera et al. [52]

are formed by direct action of ultrasounds and changes occurring in the chemical composition of the inulin-type fructans. The direct use has been suggested for inulin depolymerization, while indirect sonication would be more appropriate to extract natural fructans [18].

4.4 ANALYTICAL METHODS

Fructans analysis is fundamental because it provides basic information about these compounds mechanism of action which is dependent on its chemical structure. Fructans are commonly found as complex mixtures of carbohydrates with different DP, monomer composition and glycosidic linkages, for this reason, separation of these mixtures is not easy, because of structural and weight similarity of carbohydrates. Furthermore, their identification is hampered by lack of commercial standards [18, 40] Also, in the case of agavins, understanding properties and applications requires knowledge about sources of where they come from as well as its molecular structure [15]. Hence, analytical fructan characterization is relevant and includes different chromatographic techniques such as thin layer chromatography (TLC), high performance anion exchange chromatography combined with pulse amperometric detection (HPAEC-PAD), besides infrared spectroscopy, nuclear magnetic resonance (NMR) and matrix-assisted laser desorption–time-of-flight–mass spectrometry (MALDI-TOF-MS) among the most commonly used, nevertheless, use of these techniques is limited by technical, practical or economic reasons, such as equipment cost [15]. Another technique called fluorophore-assisted carbohydrate electrophoresis (FACE) was developed and described as sensitive, simple, and versatile but is rarely been used to analyze fructans in plants and has not been widely adopted by researchers [41].

4.4.1 INFRARED SPECTROSCOPY

Near-infrared spectroscopy (NIRS) is a method widely used for qualitative and quantitative analysis and is being used in chemical, pharmaceutical, and agro-food industries, because it is fast, simple, cheap, and non-destructive with nearly no sample preparation required [42]. One of the strengths of NIR spectroscopy is that it permits to measure several constituents at the same time. Shetty and Gislum [43] used NIR combined with chemometrics to quantify fructan concentration in grass species. They found that is possible a rapid quantification of fructans by this technique. In addition, mid-infrared spectroscopy (MIR) has been used because is a non-destructive technique and can be employed for the elucidation of molecular structures based on their functional groups [15].

4.4.2 CHROMATOGRAPHIC METHODS

Chromatography analysis of fructans started in the 1950s. Bacon [44] used chromatography on cellulose powder to quantify oligosaccharides produced by action of a yeast invertase on sucrose. While, Edelman, and Jefford [45] employed a similar method to separate 1-kestose, and other fructans. Then, Darbyshire and Henry [46] tested the same method to separate different short chain FOS ranging from 1-kestose (DP 3) to DP 6 but did not separate higher than DP 7. Rutherford and Whittle [47] separated different FOS by descending paper chromatography, but these FOS were not identified and their chemical structure not elucidated.

4.4.2.1 (THIN LAYER CHROMATOGRAPHY) TLC

TLC is a rapid, simple, and low-cost method. Compared with another techniques, this method has different advantages because it requires a simple preparation of the sample, it offers possibility of simultaneous analysis of many samples and also possibility of analysis of crude samples with a minimal preparation [48]. TLC is a technique that can resolve FOS of DP level and composition of fructans in plants. Pilon-Smits et al. [49] used TLC for fructan analysis, they separated the extracted soluble carbohydrates and used 90% acetone, and then, fructans separation was visualized with urea spray at 80°C. Reiffová and Nemcová [48] analyzed FOS as feed additives using glucose, sucrose, and fructose as standards solutions for the analysis, they pre-treated silica gel layers with 0.02 M sodium acetate, then plates were dried out at 50°C for 5 min, then, samples were applied. Layers were developed with butanol-ethanol-water as mobile phase. Spots on chromatograms were detected with a mixture of diphenylamine-aniline-phosphoric acid in acetone.

Mancilla-Margalli and López [19] compared between *Agave* and *Dasylirion* species grown in different regions of Mexico using TLC for fructan analysis, samples were applied to silica gel TLC plates with aluminum support, plates were developed with a butanol-propanol-water mixture and carbohydrate spots were visualized with aniline-diphenylamine-phosphoric acid reagent in acetone using the method of Anderson, et al [50]. Urías-Silvas and López [20] characterized *Agave* spp. and *Dasylirion* sp. fructans using silica gel plates in a saturated TLC-camera using propanol-water-butanol as mobile phase and diphenylamine-aniline-phosphoric acid as revealing reagent. While, García-Curbelo et al. [51] characterized fructans present in *A. fourcroydes* from Cuba by comparison with commercial fructans and a standard mixture with 1-kestose, nystose, kestopentaose, fructose, glucose, and sucrose, using a system of solvents of butanol-propanol-water. TLC-plates were dried and then were sprinkled with aniline-diphenylamine-phosphoric acid in an acetone base. An example of TLC separation of frutans (glucose, fructose, sucrose, 1-kestose, nystose, and fructofuranosilnystose is showed (Figure 4.2).

González-Herrera et al. [52] determined the effect of prebiotics (inulin, oligofructose, and agave fructans) on physicochemical and sensorial characteristics of a dehydrated apple matrix. They analyzed fructans with TLC using a mixture of glucose, fructose, sucrose, 1-kestose, 1-nystose,

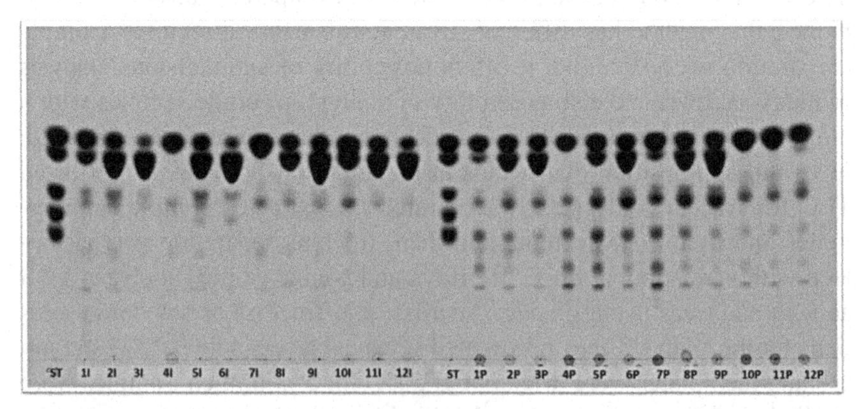

FIGURE 4.2 TLC separation of fructans from Agave sap samples. Lanes 1 and 13 (standard (glucose, fructose, sacarose, 1-kestose, 1-nystose, 1-fructofuranosilnystose (DP5)), lanes 2–12 (Agave sap samples, lanes 14–26).

and 1-fructofuranosylnystose. They applied samples on a silica gel with aluminum support plates, these plates were developed in a propanol-water-butanol solvent system and for fructans visualization, plates were sprayed with an aniline-diphenylamine-phosphoric acid in acetone reagent. Nevertheless, TLC shows a limited application, because it has low resolution and sensitivity when it is used for quantitative purposes, and furthermore could not be considered as a reliable technique for separation of high DP (>5) FOS and fructans [15, 41].

4.4.2.2 HPLC (HIGH-PERFORMANCE LIQUID CHROMATOGRAPHY) AND HPAEC-PAD (HIGH-PERFORMANCE ANION EXCHANGE CHROMATOGRAPHY WITH PULSED AMPEROMETRIC DETECTION)

In plants, fructans are generally found as poly-disperse molecules with different DP, although they are extracted from the same tissue, they have distinct fructosyl-linkages [2]. The HPLC in particular HPAEC-PAD is the most widely used method for fructan quantification, fructan distribution in a sample and is capable to provide an excellent separation of inulin-type fructans [53, 54]. Mabel et al. [55] analyzed FOS produced by a fructosyl transferase from *Aspergillus oryzae* MTCC 5154 by HPLC (LC–6A, Shimadzu) with a refractive index detector. In this case was used a polar bonded-phase column (Exsil NH2, 4.6 mm 25 cm, 5 μm), and a mixture 80:20 acetonitrile–water at a flow rate of 1.0 mL/min as mobile phase. The resulting solution was composed by glucose, sucrose, and FOS. Ortíz-Basurto et al. [56] analyzed FOS in agave sap "aguamiel" of *A. mapisaga,* using a Dionex BioLC chromatography system with PAD, with a CarboPac PA–100 column (0.4 × 25 cm). Fructans were eluted with a sodium acetate gradient in 100 Mm NaOH using 1-kestose, nystose, and inulin from Jerusalem artichoke and chicory as standards. While, Ravenscroft et al. [57] isolated fructans from *A. americana* and analyzed them by HPAEC-PAD, these authors found that fructans were composed of glucose and fructose with a DP range from 6 to 50.

Arrizon et al. [58] analyzed the composition of water-soluble carbohydrates and fructan structures from *A. tequilana.* They used Dionex BioLC50 system using an analytical CarboPac PA–100 column (4 × 250 mm). Column temperature was 35°C, and a sodium acetate gradient in 150

Mm NaOH was used at a flow rate of 1 mL min^{-1}. The elution program consisted of 6 mM sodium acetate (0–10 min), 6–500 mM (10–190 min), and 6 mM (190–200 min) using inulin from Dahlia tubers, glucose, fructose, sucrose, 1-kestose and nystose as standards. In this study, authors found that HPAEC-PAD was more useful for the analysis of low molecular weight carbohydrate than other techniques. Liu et al. [53] used a similar method with modifications. First, they made a single extraction step, selective precipitations of fructans by acetone, acid hydrolysis of precipitate and a single short HPLC run with a Dionex ED40 electrochemical detector working in pulsed amperometric mode using a gold working electrode and a combined pH-Ag/AgCl reference electrode. They found that this method is efficient and appropriate for screening of many samples. Longland et al. [59] determined fructans from pasture grasses with HPLC. Praznik et al. [60] isolated and fractionated FOS from *A. tequilana* by 2D preparative chromatography (SEC and HPLC).

Velázquez-Martínez et al. [24] analyzed the fructan profile of *A. angustifolia* Haw with HPAEC-PAD and found a branched fructan structure. On the other hand, González-Herrera et al. [52] used HPAED-PAD to determinate fructans profiles of commercial inulin, extracts of dehydrated apple matrix with or without inulin, using an ion chromatograph Dionex ICS–3000 with a CarboPac-PA–100 (4 × 250 mm) and a gradient of sodium acetate in NaOH. Nonetheless, the direct quantification of fructans by these methods is complicated because of lack of standards and poor resolution of high DP polymers, principally in grass species.

4.4.3 MALDI-TOF-MS (MATRIX-ASSISTED LASER DESORPTION TIME-OF-FLIGHT MASS SPECTROMETRY)

This method is used for assessment of spatial distribution of sugars. The choice of this method depends largely on sample nature, is an ideal technique for analysis of complex mixtures. Stahl et al. [61] analyzed fructan from *Dahlia variabilis. [2, 7], Romero-López et al. [62]* determined DP of fructans from *A. tequilana, A. salmiana,* and *A. atrovirens*. Also, Evans et al. [63] used this method to determinate DP of chicory and rye-grass. Nonetheless, unfortunately, does not provide complete structural information.

4.4.4 NMR (NUCLEAR MAGNETIC RESONANCE)

NMR of ^{13}C and ^{1}H was developed to investigate the anomeric form, ring structure, linkage oligomers and polymers containing D-fructose and β-D-fructofuranosyl groups or residues, but now is a sophisticated analytical tool used for complete fructan structural determination without sample preparation. The complex fructan structures was observed for *Agave* fructans using others analytical techniques and then confirmed by NMR [2]. Cérantola et al. [64] used this method for the characterization of inulin-type fructooligosaccharides from *Matricaria maritime* (L.). Also Ravenscroft et al. [57] determined an average DP of 14 by ^{1}H NMR of *A. americana*. Mabel et al. [55] identified oligomers present in a mixture of fructooligosaccharides by this method. Praznik et al. [60] analyzed FOS from leaves and stem of *A. tequilana.*

4.4.5 FACE (FLUOROPHORE-ASSISTED CARBOHYDRATE ELECTROPHORESIS)

FACE was described as a simple, sensitive, and versatile method for analysis of mono and oligosaccharides, FOS, and fructans and became an important technique in glycobiology. The separation of fluorescently-labeled carbohydrates is performed on polyacrylamide gels and a charge-coupled device camera used to detect and quantify the products. Jackson [65] introduced this method to separate mono and oligosaccharides and then others authors used this technique [66–69]. Although, this technique rarely has been used for analysis of FOS and fructans in plants, but it has been used to analyze other carbohydrate oligomers. Nevertheless, it has not been widely adopted by researchers [41].

4.5 FUTURE PERSPECTIVES

Recent research findings have shown health-promoting effects of agave fructans. As it was mentioned before, it has been found that agave fructans reduce the food intake, reduce glucose and triglycerides in blood, increase mineral absorption and reduce gastrointestinal upset. Most of the research has been evaluated in mice, but it is necessary to increase the research about the prebiotic effect of agave fructans and other health benefits in humans. Furthermore, agave fructans may improve food sensorial characteristics, so this and the prebiotic effect demonstrate its potential to be used as a functional ingredient in food. Nevertheless, although, agave fructans possess all of these benefits, it still requiring certification fulfillment [4].

Despite the numerous analytical techniques, the analysis of fructans still remains complex due to their structural complexity. An alternative is that fructans are best measured after their hydrolysis to D-fructose and D-glucose. It has been suggested to remove sucrose, D-glucose, and D-fructose by hydrolyzing sucrose with crystalline yeast invertase but this enzyme also hydrolyses lower polymerized fructans and FOS [70]. An alternative approach is using HPLC to analyze extracts of samples either untreated, or treated with amyloglucosidase (EC 3.2.1.3) or amyloglucosidase along with inulinase (EC 3.2.1.7, fructanase) [71]. By measuring sucrose, D-fructose, and D-glucose in various samples before and after the enzymatic hydrolysis, free D-glucose and D-fructose, sucrose, starch, and fructan can be estimated with appropriate calculation. Nonetheless, this method is not recommended for use on samples containing high levels of D-glucose, D-fructose, sucrose or maltose as these contribute to the sample blank absorbance. The next table 4.2 shows a comparison between the different analytical techniques for the analysis of agave fructans.

4.6 CONCLUSIONS

Agave fructans analysis is very important because it provides information about these potential ingredients in food industries and other applications. Nonetheless, none of the agave products has been classified as prebiotics because it is necessary to develop *in vitro* and *in vivo* research. In addition, it is needed to increase the knowledge of properties and composition of products derived from the wide biodiversity of *Agave* plants. Some procedures for measurement of FOS and fructans levels in plants, crops, and food products were described. Chromatography-based techniques are the most used methods specially HPAEC-PAD. All techniques show limitation but each method had advantages. It is important to highlight that is necessary to know about the sample that will be analyzed and then choose the best and more reliable technique to elucidate the structure and function of fructans.

TABLE 4.2 Comparison of Analytical Techniques for Agave Fructans

Technique	Function	Advantages	Disadvantages	Reference
Infrared spectroscopy	Elucidate structures based on their functional groups	– Rapid, simple, and low-cost. – Non-destructive method. – The sample not required preparation.	–	[42]
TLC	Elucidate different degrees of polymerization.	– Rapid, simple, and low-cost. – Simple preparation of the sample. – Possibility of simultaneous analysis.	– Not reliable and reproducible with high DP (>5) fructans.	[41]
HPAEC-PAD	Fructan quantification.	– Provide an excellent separation of inulin-type fructans.	– Lack of standards. – Poor resolution of high DP fructans.	[53, 54]
MALDI-TOF-MS	Used for assessment of spatial distribution of sugars.	– Ideal technique for analysis of complex mixtures.	– Not provide complete structural information.	[15]
NMR	Used to elucidate complete fructan structural information.	– Preparation of sample is not required.	–	[15]
FACE	Used to detect and quantify.	– Simple, sensitive, and versatile for fructans.	– This technique has rarely been used for the analysis of fructans in plants.	[41]

ACKNOWLEDGMENTS

S.E.G. wants to thank the Mexican National Council of Science and Technology (CONACYT) for the financial support during her postgraduate studies.

KEYWORDS

- **chromatographic methods**
- **Graminans**
- **inulin**
- **levans**
- **prebiotic**

REFERENCES

1. Mellado-Mojica, E., & López, M. G., (2012). Fructan metabolism in *A. tequilana* weber blue variety along its developmental cycle in the field. *J. Agric. Food Chem.*, *60*(47), 11704–11713.
2. López, M. G., & Mancilla-Margalli, N. A., (2007). The nature of fructooligosaccharides in Agave plants. In: Norio, S., Noureddine, B., & Shuichi, O., (eds.), *Recent Advances in Fructooligosaccharides Research* (pp. 47–67).
3. Michel-Cuello, C., Juárez-Flores, B. I., Aguirre-Rivera, J. R., & Pinos-Rodríguez, J. M., (2008). Quantitative characterization of nonstructural carbohydrates of mezcal agave (*Agave salmiana* Otto ex Salm-Dick). *J. Agric. Food Chem.*, *56*(14), 5753–5757.
4. Carranza, C. O., Fernandez, A. Á., Bustillo Armendáriz, G. R., & López-Munguía, A., (2015). Chapter 15 - Processing of fructans and oligosaccharides from agave plants A2 - preedy, victor. In: *Processing and Impact on Active Components in Food* (pp. 121–129). San Diego, Academic Press.
5. Hendry, G. A. F., & Wallace, R. K., (1993). The origin, distribution and evolutionary significance of fructans. In: Suzuki, M., & Chatterton, N. J., (eds.), *Science and Technology of Fructans* (pp. 119–136). Taylor & Francis.
6. Franco-Robles, E., & López, M. G., (2015). Implication of fructans in health: Immunomodulatory and antioxidant mechanisms. *Sci. World J.*, 289267.
7. Michel-Cuello, C., Ortíz-Cerda, I., Moreno-Vilet, L., Grajales-Lagunes, A., Moscosa-Santillán, M., Bonnin, J. et al., (2012). Study of enzymatic hydrolysis of fructans from *Agave salmiana* characterization and kinetic assessment. *Sci. World J.*, 10.

8. López, M. G., Mancilla-Margalli, N. A., & Mendoza-Díaz, G., (2003). Molecular structures of fructans from *Agave tequilana* Weber var. azul. *J. Agric. Food Chem.*, *51*(27), 7835–7840.
9. Frehner, M., Keller, F., & Wiemken, A., (1984). Localization of fructan metabolism in the vacuoles isolated from protoplasts of jerusalem artichoke tubers (*Helianthus tuberosus* L.). *J. Plant Physiol.*, *116*(3), 197–208.
10. Ritsema, T., & Smeekens, S., (2003). Fructans: Beneficial for plants and humans. *Curr. Opin. Plant Biol.*, *6*(3), 223–230.
11. Wagner, W., Keller, F., & Wiemken, A., (1983). Fructan metabolism in cereals: Induction in leaves and compartmentation in protoplasts and vacuoles. *Zeitschrift für Pflanzenphysiologie.*, *112*(4), 359–372.
12. Hendry, G. A. F., (1993). Evolutionary origins and natural functions of fructans - a climatological, biogeographic and mechanistic appraisal. *New Phytol.*, *123*(1), 3–14.
13. Vereyken, I. J., Chupin, V., Demel, R. A., Smeekens, S. C. M., & De Kruijff, B., (2001). Fructans insert between the headgroups of phospholipids. *Biochim. Biophys. Acta, Biomembr.*, *1510*(1–2), 307–320.
14. Livingston, D. P., & Henson, C. A., (1998). Apoplastic sugars, fructans, fructan exohydrolase, and invertase in winter oat: Responses to second-phase cold hardening. *Plant Physiol.*, *116*(1), 403–408.
15. López, M. G., Huazano-García, A., García-Pérez, M. C., & García-Vieyra, M. I., (2014). Agave fiber structure complexity and its impact on health. In: Benkeblia, N., (ed.), *Polysaccharides* (pp. 45–74). CRC Press.
16. Hincha, D. K., Zuther, E., Hellwege, E. M., & Heyer, A. G., (2002). Specific effects of fructo- and gluco-oligosaccharides in the preservation of liposomes during drying. *Glycobiology*, *12*(2), 103–110.
17. González-Cruz, L., Jaramillo-Flores, M. E., Bernardino-Nicanor, A., & Mora-Escobedo, R., (2011). Influence of plant age on fructan content and fructosyltranserase activity in *Agave atrovirens* Karw leaves. *Afr. J. Biotechnol.*, *10*(71), 15911.
18. Apolinário, A. C., De Lima Damasceno, B. P. G., De Macêdo, B. N. E., Pessoa, A., Converti, A., & Da Silva, J. A., (2014). Inulin-type fructans: A review on different aspects of biochemical and pharmaceutical technology. *Carbohydr. Polym.*, *101*, 368–378.
19. Mancilla-Margalli, N. A., & López, M. G., (2006). Water-soluble carbohydrates and fructan structure patterns from *Agave* and *Dasylirion* species. *J. Agric. Food Chem.*, *54*(20), 7832–7839.
20. Urías-Silvas, J. E., & López, M. G., (2009). *Agave* spp. and *Dasylirion* sp. Fructans as a potential novel source of prebiotics. *Dyn. Biochem. Biotechnol. Mol. Biol.*, *3*, 59–65.
21. Gibson, G. R., Probert, H. M., Loo, J. V., Rastall, R. A., & Roberfroid, M. B., (2004). Dietary modulation of the human colonic microbiota: Updating the concept of prebiotics. *Nutr. Res. Rev.*, *17*(2), 259–275.
22. Gibson, G. R., & Roberfroid, M. B., (1995). Dietary modulation of the human colonic microbiota: Introducing the concept of prebiotics. *J. Nutr.*, *125*(6), 1401–1412.
23. Santiago-García, P. A., & López, M. G., (2009). Prebiotic effect of agave fructans and mixtures of different degrees of polimerization from *Agave angustifolia* Haw. *Dyn. Biochem. Biotechnol. Mol. Biol.*, *3*, 52–58.

24. Velázquez-Martínez, J. R., González-Cervantes, R. M., Hernández-Gallegos, M. A., Mendiola, R. C., Aparicio, A. R., & Ocampo, M. L., (2014). Prebiotic potential of *Agave angustifolia* Haw fructans with different degrees of polymerization. *Mol.*, *19*(8), 12660–12675.
25. Urías-Silvas, J. E., Cani, P. D., Delmee, E., Neyrinck, A., López, M. G., & Delzenne, N. M., (2008). Physiological effects of dietary fructans extracted from *Agave tequilana* Gto. and *Dasylirion* spp. *Br. J. Nutr.*, *99*(2), 254–261.
26. Santiago-García, P. A., & López, M. G., (2014). Agavins from *Agave angustifolia* and *Agave potatorum* affect food intake, body weight gain and satiety-related hormones (GLP–1 and ghrelin) in mice. *Food Funct.*, *5*(12), 3311–3319.
27. García-Vieyra, M. I., Del Real, A., & López, M. G., (2014). Agave fructans: Their effect on mineral absorption and bone mineral content. *J. Med. Food*, *17*(11), 1247–1255.
28. García-Curbelo, Y., Bocourt, R., Savon, L. L., García-Vieyra, M. I., & López, M. G., (2015). Prebiotic effect of Agave fourcroydes fructans: an animal model. *Food Funct.*, *6*(9), 3177–3182.
29. Huazano-García, A., & López, M. G., (2015). Agavins reverse the metabolic disorders in overweight mice through the increment of short chain fatty acids and hormones. *Food Funct.*, *6*(12), 3720–3727.
30. Franco-Robles, E., & López, M., (2016). Agavins increase neurotrophic factors and decrease oxidative stress in the brains of high-fat diet-induced obese mice. *Mol.*, *21*(8), 998.
31. Holscher, H. D., Doligale, J. L., Bauer, L. L., Gourineni, V., Pelkman, C. L., Fahey, G. C. et al., (2014). Gastrointestinal tolerance and utilization of agave inulin by healthy adults. *Food Funct.*, *5*(6), 1142–1149.
32. López-Velázquez, G., Parra-Ortíz, M., Mora Ide, L., García-Torres, I., Enriquez-Flores, S., Alcantara-Ortigoza, M. A. et al., (2015). Effects of fructans from Mexican agave in newborns fed with infant formula: A randomized controlled trial. *Nutr.*, *7*(11), 8939–8951.
33. Zamora-Gasga, V. M., Bello-Pérez, L. A., Ortíz-Basurto, R. I., Tovar, J., & Sáyago-Ayerdi, S. G., (2014). Granola bars prepared with *Agave tequilana* ingredients: Chemical composition and in vitro starch hydrolysis. *LWT-Food Sci. Technol.*, *56*(2), 309–314.
34. Zamora-Gasga, V. M., Loarca-Piña, G., Vázquez-Landaverde, P. A., Ortíz-Basurto, R. I., Tovar, J., & Sáyago-Ayerdi, S. G., (2015). *In vitro* colonic fermentation of food ingredients isolated from *Agave tequilana* Weber var. azul applied on granola bars. *LWT-Food Sci. Technol.*, *60*(2, Part 1), 766–772.
35. Crispín-Isidro, G., Lobato-Calleros, C., Espinosa-Andrews, H., Álvarez-Ramírez, J., & Vernon-Carter, E. J., (2015). Effect of inulin and agave fructans addition on the rheological, microstructural and sensory properties of reduced-fat stirred yogurt. *LWT-Food Sci. Technol.*, *62*(1, Part 2), 438–444.
36. Kim, Y., Faqih, M. N., & Wang, S. S., (2001). Factors affecting gel formation of inulin. *Carbohydr. Polym.*, *46*(2), 135–145.

37. González-Cruz, L., E, J. F. M., Bernardino-Nicanor, A., & Mora-Escobedo, R., (2012). Influence of the harvest age on fructan content and fructosyltransferase activity in *Agave atrovirens* karw pine. *Int. J. Plant Physiol. Biochem.*, *4*(5), 110–119.
38. Prosky, L., & Hoebregs, H., (1999). Methods to determine food inulin and oligofructose. *J. Nutr.*, *129*(7), 1418–1423.
39. Dalonso, N., Ignowski, E., Monteiro, C. M. A., Gelsleichter, M., Wagner, T. M., Silveira, M. L. L. et al., (2009). Extração e caracterização de carboidratos presentes no alho (*Allium sativum* L.): Proposta de metodologia alternativa. *Food Sci. Technol.*, *29*, 793–797.
40. Brokl, M., Hernández-Hernández, O., Soria, A. C., & Sanz, M. L., (2011). Evaluation of different operation modes of high performance liquid chromatography for the analysis of complex mixtures of neutral oligosaccharides. *J. Chromatogr. A*, *1218*(42), 7697–7703.
41. Benkeblia, N., (2013). Fructooligosaccharides and fructans analysis in plants and food crops. *J. Chromatogr. A*, *1313*, 54–61.
42. Alessandrini, L., Romani, S., Pinnavaia, G., & Rosa, M. D., (2008). Near infrared spectroscopy: An analytical tool to predict coffee roasting degree. *Anal. Chim. Acta*, *625*(1), 95–102.
43. Shetty, N., & Gislum, R., (2011). Quantification of fructan concentration in grasses using NIR spectroscopy and PLSR. *Field Crops Research*, *120*(1), 31–37.
44. Bacon, J. S. D., (1954). The oligosaccharides produced by the action of yeast invertase preparations on sucrose. *Biochem. J.*, *57*(2), 320–328.
45. Edelman, J., & Jefford, T., (1964). The metabolism of fructose polymers in plants. *Biochem. J.*, *93*(1), 148–161.
46. Darbyshire, B., & Henry, R. J., (1978). The distribution of fructans in onions. *New Phytol.*, *81*(1), 29–34.
47. Rutherford, P. P., & Whittle, R., (1982). The carbohydrate composition of onions during long term cold storage. *J. Hortic. Sci.*, *57*(3), 349–356.
48. Reiffová, K., & Nemcová, R., (2006). Thin-layer chromatography analysis of fructooligosaccharides in biological samples. *J. Chromatogr. A*, *1110*(1–2), 214–221.
49. Pilon-Smits, E. A. H., Ebskamp, M. J. M., Paul, M. J., Jeuken, M. J. W., Weisbeek, P. J., & Smeekens, S. C. M., (1995). Improved performance of transgenic fructan-accumulating tobacco under drought stress. *Plant Physiol.*, *107*(1), 125–130.
50. Anderson, K., Li, S. C., & Li, Y. T., (2000). Diphenylamine-aniline–phosphoric acid reagent, a versatile spray reagent for revealing glycoconjugates on thin-layer chromatography plates. *Anal. Biochem.*, *287*(2), 337–339.
51. García-Curbelo, Y., López, M. G., Bocourt, R., Collado, E., Albelo, N., & Nuñez, O., (2016). Structural characterization of fructans from *Agave fourcroydes* (Lem.) with potential as prebiotic. *Cuban J. Agric. Sci.*, *49*(1).
52. González-Herrera, S. M., Rutiaga-Quiñones, O. M., Aguilar, C. N., Ochoa-Martínez, L. A., Contreras-Esquivel, J. C., López, M. G. et al., (2016). Dehydrated apple matrix supplemented with agave fructans, inulin, and oligofructose. *LWT-Food Sci. Technol.*, *65*, 1059–1065.

53. Liu, Z., Mouradov, A., Smith, K. F., & Spangenberg, G., (2011). An improved method for quantitative analysis of total fructans in plant tissues. *Anal. Biochem., 418*(2), 253–259.
54. Legnani, G., & Miller, W. B., (2001). Short photoperiods induce fructan accumulation and tuberous root development in Dahlia seedlings. *New Phytol., 149*(3), 449–454.
55. Mabel, M. J., Sangeetha, P. T., Platel, K., Srinivasan, K., & Prapulla, S. G., (2008). Physicochemical characterization of fructooligosaccharides and evaluation of their suitability as a potential sweetener for diabetics. *Carbohydr. Res., 343* (1), 56–66.
56. Ortíz-Basurto, R. I., Pourcelly, G., Doco, T., Williams, P., Dornier, M., & Belleville, M. P., (2008). Analysis of the main components of the aguamiel produced by the maguey-pulquero (*Agave mapisaga*) throughout the harvest period. *J. Agric. Food Chem., 56*(10), 3682–3687.
57. Ravenscroft, N., Cescutti, P., Hearshaw, M. A., Ramsout, R., Rizzo, R., & Timme, E. M., (2009). Structural analysis of fructans from *Agave americana* grown in South Africa for spirit production. *J. Agric. Food Chem., 57*(10), 3995–4003.
58. Arrizon, J., Morel, S., Gschaedler, A., & Monsan, P., (2010). Comparison of the water-soluble carbohydrate composition and fructan structures of *Agave tequilana* plants of different ages. *Food Chem., 122*(1), 123–130.
59. Longland, A. C., Dhanoa, M. S., & Harris, P. A., (2012). Comparison of a colorimetric and a high-performance liquid chromatography method for the determination of fructan in pasture grasses for horses. *J. Sci. Food Agric., 92*(9), 1878–1885.
60. Praznik, W., Löppert, R., Cruz Rubio, J. M., Zangger, K., & Huber, A., (2013). Structure of fructo-oligosaccharides from leaves and stem of *Agave tequilana* Weber, var. azul. *Carbohydr. Res., 381*, 64–73.
61. Stahl, B., Linos, A., Karas, M., Hillenkamp, F., & Steup, M., (1997). Analysis of fructans from higher plants by matrix-assisted laser desorption/ionization mass spectrometry. *Anal. Biochem., 246*(2), 195–204.
62. Romero-López, M. R., Osorio-Díaz, P., Flores-Morales, A., Robledo, N., & Mora-Escobedo, R., (2015). Chemical composition, antioxidant capacity and prebiotic effect of aguamiel (*Agave atrovirens*) during in vitro fermentation. *Revista Mexicana de Ingeniería Química., 14*, 281–292.
63. Evans, M., Gallagher, J. A., Ratcliffe, I., & Williams, P. A., (2016). Determination of the degree of polymerisation of fructans from ryegrass and chicory using MALDI-TOF mass spectrometry and gel permeation chromatography coupled to multiangle laser light scattering. *Food Hydrocolloids., 53*, 155–162.
64. Cérantola, S., Kervarec, N., Pichon, R., Magné, C., Bessieres, M. A., & Deslandes, E., (2004). NMR characterisation of inulin-type fructooligosaccharides as the major water-soluble carbohydrates from *Matricaria maritima* (L.). *Carbohydr. Res., 339*(14), 2445–2449.
65. Jackson, P., (1990). The use of polyacrylamide-gel electrophoresis for the high-resolution separation of reducing saccharides labelled with the fluorophore 8-aminonaphthalene–1,3,6-trisulphonic acid. Detection of picomolar quantities by an imaging system based on a cooled charge-coupled device. *Biochem. J., 270*(3), 705–713.

66. Harvey, D. J., (2011). Derivatization of carbohydrates for analysis by chromatography, electrophoresis and mass spectrometry. *J. Chromatogr. B*, *879*(17–18), 1196–1225.
67. Starr, C. M., Irene Masada, R., Hague, C., Skop, E., & Klock, J. C., (1996). Fluorophore-assisted carbohydrate electrophoresis in the separation, analysis, and sequencing of carbohydrates. *J. Chromatogr. A*, *720*(1), 295–321.
68. Hu, G. F., (1995). Analytical biotechnology fluorophore-assisted carbohydrate electrophoresis technology and applications. *J. Chromatogr. A*, *705*(1), 89–103.
69. O'Shea, M. G., Samuel, M. S., Konik, C. M., & Morell, M. K., (1998). Fluorophore-assisted carbohydrate electrophoresis (FACE) of oligosaccharides: Efficiency of labelling and high-resolution separation. *Carbohydr. Res.*, *307*(1–2), 1–12.
70. Pontis, H. G., (1990). 10 - Fructans. In: Dey, P. M., (ed.), *Methods in Plant Biochemistry* (Vol. 2, pp. 353–369). Academic Press.
71. Quemener, B., Thibault, J. F., & Coussement, P., (1994). Determination of inulin and oligofructose in food products, and integration in the AOAC method for measurement of total dietary fibre. *LWT-Food Sci. Technol.*, *27*(2), 125–132.
72. Ávila-Fernández, Á., Galicia-Lagunas, N., Rodríguez-Alegría, M. E., Olvera, C., & López-Munguía, A., (2011). Production of functional oligosaccharides through limited acid hydrolysis of agave fructans. *Food Chem.*, *129*(2), 380–386.
73. Espinosa-Andrews, H., & Urías-Silvas, J. E., (2012). Thermal properties of agave fructans (*Agave tequilana* Weber var. Azul). *Carbohydr. Polym.*, *87*(4), 2671–2676.
74. Montañez-Soto, J. L., Venegas-González, J., Ceja-Torres, L. F., Castellanos-Pérez, N., & Yañez-Fernández, J., (2014). Fractional extraction of the fructans contained in the *agave tequilana* weber blue head based on their average degree of polimerization. *Adv. Bio Res.*, *5*(3), 107–113.
75. Moreno-Vilet, L., Camacho-Ruiz, R. M. & Portales-Pérez, D. P. (2015). Prebiotic Agave Fructans and Immune Aspects. *In:* Watson, R. R. & Preedy, V. R. (eds.). *Probiotics, Prebiotics, and Synbiotics: Bioactive Foods in Health Promotion.* First edition: Academic Press, Elsevier Inc.

CHAPTER 5

AGUAMIEL: TRADITIONAL MEXICAN PRODUCT AS A NUTRITIONAL ALTERNATIVE

BRIAN PICAZO, ADRIANA CAROLINA FLORES GALLEGOS, and CRISTÓBAL N. AGUILAR

Bioprocesses Group, Food Research Department, Faculty of Chemistry, Autonomous University of Coahuila, Saltillo, 25280, Coahuila, Mexico, E-mail: cristobal.aguilar@uadec.edu.mx

ABSTRACT

Aguamiel is a natural product obtained from *Agave* plants. In Mexico, over 200 *Agave* species can be found across all the territory but aguamiel it is not produced in all species. Some of the species producers of aguamiel are *Agave salmiana*, *Agave atrovirens* or *Agave mapisaga* to name some of them. Aguamiel is a liquid product and can be used as a beverage or as a raw material to produce some commercial products like "pulque" or "pulque's bread"; it takes a long time to be produced by the plant and recollection is done manually. Sugars are the main compounds in aguamiel, including glucose, sucrose, fructose, and a quantity of short-chain fructo-oligosaccharides; also, it contains some minerals like Mg^{+2} and Ca^{+2}, saponins and a few amino acids. The presence of fructo-oligosaccharides brings some health benefits to this sap; for example, acts as prebiotic or enhances mineral absorption of mineral in the colon. Due to the beneficial properties and composition that it presents, aguamiel may be used as a new alternative to improve nutraceutical of functional foods.

5.1 INTRODUCTION

Mexico has different kind of natural resources, which are a good economic source to the people producing them; *Agave* production is one of them. In Mexico territory above 200 different species are found, representing 75% of all species around the world [1]. However, *Agave* plants species distribution and production varies within Mexico's territory. Some states have more production of a different variety as it is shown in Table 5.1.

TABLE 5.1 Principal States Where Subgenres of *Agave* Can Be Found

Subgenre	Species	Zones	Principal states	Reference
Agave	*A. salmiana*	Sierra Madre Oriental	Baja California	[2]
	A. atrovirens	Sierra Madre Occidental	Durango	[3]
	A. angustifolia	Southern Highlands	Sonora	[4]
	A. tequilana	Sierra Madre del Sur	Chihuahua	[2]
			Coahuila	[5]
			Tamaulipas	
			Guanajuato	
			Querétaro	
			Jalisco	
Littaea	*A. celsii albicans*	Center and East of Mexico	Hidalgo	[2]
	A. xylonacantha		Querétaro	[6]
	A. difformis		Puebla	
	A. striata		Sonora	
			Durango	

Mesoamerican cultures used to use *Agave* products as a medicinal alternative when they get sick without knowing any of their medicinal or nutritional properties. The different cultures used and consumed many parts of the *Agave* plant, mainly the cooked heart and leaves. The mainly *Agave* products that can be extracted from this plant are its sap (aguamiel) and pulque, which is the result of spontaneous fermentation of agave sap. Pulque is consumed as a traditional alcoholic beverage or used as raw material to produce "pulque bread." However, because of its high sugar content, aguamiel can also be used for the production of fructans and agave syrups [7, 8]. On the other hand, there are other products from

Agave plants very popular nowadays around the world, tequila and mezcal, both alcoholic beverages produced after a distillation process [5, 9] from different plant species.

5.2 AGUAMIEL PRODUCTION AND EXTRACTION

Aguamiel is naturally produced by different *Agave* species. Each one must and produce and exudate that will be collected. Some of the *Agave* species used to produce aguamiel are *Agave americana, A. atrovirens, A. ferox, A. mapisaga* or *A. salmiana*. In Coahuila, the aguamiel that is marketed is produced from plants of *A. salmiana* and *A. atrovirens* [Narváez-Zapata and Sánchez-Teyer 2010, 3].

The first step in aguamiel production is to allow the *Agave* plant to grow and mature. Usually, the time it takes a plant to reach maturity is between 8 and 10 years; When the plant is ready to produce aguamiel, *Agave* stalk is observed in vertical position (Figure 5.1), and then it needs to be cut to generate a cavity in the center of the plant where the exudate will be collected (Figure 5.2); inside this cavity, aguamiel will be produced naturally by exudation [10]. For this, the cavity walls will be scraped continuously.

Once the cavity has been generated in the center of the plant, it must be covered to protect it from the aguamiel of the animals living around the plantation, such as coyotes and foxes, which may be attracted by its sweet fragrance (Figure 5.3). The plant is prepared to produce aguamiel all over four months; every time the aguamiel is collected, the cover is removed and placed again. The aguamiel collection is carried out with an "acocote," an old traditional instrument; however, alternatively the aguamiel is collected by introducing into the cavity glass vessels or cups while the collected liquid is transferred to bigger bottles (Figure 5.4) [10].

The final step of this process consists on scraping the walls of the cavity made in the *Agave* (Figure 5.5). This is done with the help of a rounded spatula in order that the walls of the cavity do not dry and facilitate the sap that exudes the plant to accumulate. The plant material that is scraped is known as bagasse and is collected and discarded. The cavity is covered again, and this process is repeated every day until the *Agave* stops producing aguamiel [10].

FIGURE 5.1 *Agave* stalks in vertical position.

The collection process is carried out for about four months approximately since the first production; however, the amount of aguamiel produced varies during this period (it generally decreases) as well as depending on each *Agave* plant and species. Some plants can produce up to five liters per day while others have a maximum production of two or three liters per day. The process of production of aguamiel takes a long time, and is done in an artisan way since special care must be taken to scrape the walls of the cavity and remove the bagasse. If this process is carried out incorrectly, the production of aguamiel can be reduced or lost altogether.

FIGURE 5.2 Generating the cavity in the center of *Agave*.

5.2.1 AGUAMIEL COMPOSITION

Aguamiel consists of different compounds, mainly sugars as Ortiz-Basurto et al. [11] reported. Sugars present in aguamiel are diverse, but glucose is found in greater quantity, followed by sucrose, fructose, and a lower

FIGURE 5.3 Cavity covered to protect the plant.

amount of fructo-oligosaccharides. Sugar concentrations in aguamiel may vary along the four seasons of the year as Enríquez Salazar [3] reported; in winter and spring it has the highest sugar content, when the weather is extreme (very cold or very hot) and the most abundant microbiota. Aguamiel also has low amounts of proteins and free amino acids that can favor the human diet; some essential amino acids found in aguamiel are aspartic acid, arginine, and alanine [11]. These main compounds are shown in Table 5.2.

Aguamiel has a great content and variety of minerals, most of which have a beneficial effect on health. Some of them are calcium, followed by iron and magnesium, and other minerals. Calcium is the mineral found in greater proportion in aguamiel of different *Agave* species. For example, *Agave salmiana* has 408 mg L^{-1} according to Tovar et al. [2008], which is

FIGURE 5.4 Collection of aguamiel.

TABLE 5.2 Compounds Present in Aguamiel From Different *Agave* Plants

Total sugars	Reducing sugars	Proteins	Lipids	Ashes	Brix degrees	Density (g/cm^3)	Reference
219.09 [c]	23.10 [c]	0.65 [c]	N/A	N/A	9.78	N/A	[3]
254.15 [c]	32.15 [c]	0.41 [c]	N/A	N/A	13.66	N/A	[3]
102.88 [c]	32.33 [c]	0.14 [c]	18.58 [c]	8.17 [c]	10.60	1.05	[12]
N/A	61.31 [d]	3.50 [d]	N/A	3.10 [d]	N/A	N/A	[13]

[a] *Agave salmiana*, [b] *Agave atrovirens*, [c] g/L, [d]% dw.

higher than the *Agave salmiana* reported by Silos-Espino et al. [2010] with 200 mg L^{-1}; meanwhile, *Agave mapisaga* reported from Ortíz-Basurto et al. [11] contains 126 mg L^{-1}. On the other hand, both *Agave salmiana* and aguamiel from *Agave atrovirens* have a high content of copper and zinc,

FIGURE 5.5 Scraping process of the cavity in *Agave*.

and a consumption of 850 mL will provide the recommended daily intake of iron and zinc.

Aguamiel from *Agave atrovirens* also has a high amount of aminoacids, containing 17 different aminoacids, 9 of these are essential aminoacids and some vitamins such as thiamine, rivoflavine, niacin, pirodixine, and vitamin C [13, 14]. Meanwhile, aguamiel from *Agave salmiana* has not yet been characterized about these compounds.

Saponins can be found in aguamiel at a high concentration compared to another sources; Leal-Díaz et al. [15] reported a high concentration of kammogenin before aguamiel gets mature, reducing its concentration almost a half in *Agave salmiana*. Also, aguamiel contains manogenin, gentrogenin, and hecogenin, but in low concentrations compared to kammogenin [15, 16]. Due all the different saponins content, aguamiel can reach 1.17 g/100 g [13].

5.2.2 HEALTH BENEFITS

Within aguamiel composition, several compounds can improve human health in many ways. Sugars are the major component in aguamiel and, within them, fructo-oligosaccarides are those that are in greater concentration. These compounds play an important role in the beneficial effects of aguamiel.

The main benefit of fructo-oligosaccharides in human health is that they can act as prebiotic compounds and favor the beneficial microbiota inside colon that use the FOS as carbon source for its growth. These beneficial microorganisms can assist in the food processing: Yen et al. [17] reported that 10 g of FOS per day increase the excretion of bifidobacteria and reduce plasma thiobarbituric acid-reactive substances. FOS also relief constipation and reduce the formation of putrefaction products in the gut, and reduce triglycerides, cholesterol, glucose, and blood pressure as reported by Hidaka et al. [18]. This agrees with the reported by Phuwamongkolwiwat et al. [19] who reported the regulation of cholesterol and triglycerides levels with increased secretion of GLP–1 and quercetin in the blood. Fructo-oligosaccharides can provide anticancer properties in colon according to Vitali et al. [20], who reported that these compounds enhance the proliferation of microorganisms in the colon which inactivate the production of toxic compounds, reduce the concentrations of pyridine and increase butyrate production, while reducing the adsorption of heterocyclic amines. Finally, increasing the population of these microorganisms in the colon favors the intestinal absorption of minerals such as Na^{+} and Mg^{+2} [21].

Due to the amino acid, mineral, and vitamin content in aguamiel, this product can provide different benefits as a nutritious food. Aguamiel vitamins are essential for cellular respiration and synthesis of different compounds as in the case of vitamin B6. Also, its diverse amino acid content and high amount of essential amino acids makes it a good source of these for human consumption. Also, kammogenin and manogenin can provide beneficial properties, as they act as antiproliferative cancer cells [16].

5.2.3 MICROORGANISMS ISOLATED FROM AGUAMIEL

As aguamiel has a high concentration of sugars, its variety of microorganisms is high; many of these microorganisms are beneficial to human health. Castro-Rodríguez et al. [22] isolated four *Leuconostoc mesenteroides* strain with probiotic activity from aguamiel of *Agave salmiana.* When comparing these strains with a commercial strain of *Lactobaccilus plantarum* v299, the four strains showed the property of adhering to the intestinal mucosa and a high antimicrobial activity. Castro-Rodríguez [23], reported isolation of a *Leuconostoc mesenteroides* strain that was compared to a strain of *Lactobacillus delbruecki* subsp. *Bulgaricus* NRRL-B–734; the *L. mesenteroides* strain showed a better adhesion to the intestinal mucosa and a better antimicrobial activity against *Salmonella typhimurium* and *Listeria monocytogenes,* both very pathogen microorganisms.

From aguamiel not only strains have been isolated to evaluate their probiotic properties, another yeast strain was also evaluated which act as a killer indicating that these yeasts can be used as starters in a fermentation processes. Estrada-Godina et al. [24] isolated some yeast strains from aguamiel including *Candida lusitaneae,* two strains of *Kluyveromyces marxianus* var. *bulgaricus* and a *Saccharomyces cerevisiae* strain. However, only the *Kluyveromyces marxianus* var. *bulgaricus* strains showed a killer activity, so only these two strains can be used in alcohol production.

5.3 AGUAMIEL APPLICATION IN DIVERSE PRODUCTS

Aguamiel can be used to produce other derivatives such as pulque, pulque bread, and agave syrup.

5.3.1 PULQUE

Pulque was known by pre-Hispanic cultures (like the Aztecs) like *metoctli, iztacoctlli* o *poliuhquioctli*. These definitions mean that this product is a wine and, possibly, the Spanish decided to use this last definition as a basis for calling this product *pulque* and thus refer that fermented pre-Hispanic drink [7].

Pulque is an acidic, viscous, and alcoholic traditional Mexican beverage produced from aguamiel [Valadez-blanco et al., 2012]. It is produced mainly in wooden or plastic containers and is the product of the spontaneous fermentation process of aguamiel. The fermentations process is a fed-batch fermentation at 25 – 50°C between 12 to 24 h and the final alcohol concentrations can vary from 3 to 6% [Valadez-blanco et al. 2012]. The main microorganisms involved in this fermentation are *Zymomonas mobilis, Leuconostoc mesenteroides* or *Lactobacillus sp.*, and sometimes also can be found involved in this process *Saccharomyces cerevisiae.* All these bacteria or fungi are naturally found in aguamiel when extracted from *Agave* plants [3, 7, 25].

5.3.2 PULQUE BREAD

Pulque's bread is a traditional product from Saltillo, Coahuila (northern Mexico) since 1925 [26]. This product is made since ancient times and it was produced after people discovered *pulque.* Pulque bread is made with pulque, brown sugar and cinnamon as main components, and is elaborated like a normal bread; in some ancient reports, this bread was salted, but later people made it into a sweet bread by adding sugar to its elaboration [Rios, 2017]. The production of this bread is still in the home industry and is sold in a few specialized stores on its production.

5.3.3 AGAVE SYRUP

Agave syrup is a high-fructose product, which is obtained after the release of fructose as a result of the hydrolysis of the fructo-oligosaccharides contained in the aguamiel [28]. The aguamiel has 74°Brix and pH can vary between 4 to 6, while fructose can reach 70% of the total sugar composition after hydrolysis, and less than 2% of sucrose [28]. However, Muñiz-Márquez et al. [12] reported an increase in sucrose concentration after applying a heat treatment to aguamiel, increasing up to 20 times the initial sucrose concentration.

After this process aguamiel, the sugar content is concentrated, so that all the beneficial attributes aguamiel are multiplied. Some compounds extracted from this concentrated product are flavonoids, polycosanols or sapogenins. These compounds can inhibit colon cancer Caco–2 and hepatic cancer HepG2 cell lines by up to 84.8 and 67.9% respectively; in addition, acetone extracts had a high antioxidant activity, these extracts could be applied in some products to increase their nutritional content [28, 29].

5.4 CONCLUSIONS

Aguamiel contains a considerable number of beneficial compounds to the health; this product can be used in many products. It can be used as an ingredient in some foods or it can be converted into many products: as a processed beverage, as a concentrate of aguamiel or another. Also, it has a high potential of strain isolation, so it can be used to obtain microorganisms that help human health or improve some biotechnological processes like fermentation or another bioprocess.

KEYWORDS

- **aguamiel**
- **biotechnological processes**
- **fermentation**

REFERENCES

1. Narváez-Zapata, J. A., & Sánchez-Teyer, L. F., (2010). Agaves as a raw material: Recent technologies and applications. *Recent Pat. Biotechnol.*, *3*(3), 185–191.
2. García-Mendoza, A. J., (2007). Los agaves de México. *Ciencias*, 087.

3. Enríquez, S. M. I., (2015). Diversidad microbiana asociada a la recolección de aguamiel de *agave atrovirens* y *agave salmiana*. MSc Dissertation, Universidad Autónoma de Coahuila. Saltillo, Coah.
4. Barraza-Morales, A., Sánchez-Teyer, F. L., Robert, M., Esqueda, M., & Gardea, A., (2006). Variabilidad genética en *Agave angustifolia* Haw. de la Sierra Sonorense, México, determinada con marcadores AFLP. *Rev. Fitotec. Mex.*, *29*(1), 1-8.
5. Bautista, J. M., Garcia, O. L., Salcedo, H. R., & Parra, N. L. A., (2001). Azúcares en agaves (*Agave tequilana* Weber) cultivados en el estado de Guanajuato. *Acta Universitaria.*, *11*(1), 33-38.
6. Rocha, M., Valera, A., & Eguiarte, L. E., (2005). Reproductive ecology of five sympatric *Agave Littaea* (*Agavaceae*) species in central Mexico. *Am. J. Bot.*, *92*(8), 1330–1341.
7. Escalante, A., Soto, D. R. L., Gutiérrez, J. E. V., Giles-Gómez, M., Bolívar, F., & López-Munguía, A., (2016). Pulque, a traditional Mexican alcoholic fermented beverage: Historical, microbiological, and technical aspects. *Front Microbiol.*, 7.
8. Robertson, I. G., & Cortés, M. O. C., (2017). Teotihuacan pottery as evidence for subsistence practices involving maguey sap. *Archaeol. Anthropol. Sci.*, *9*(1), 11–27.
9. Martínez-Gándara, A., (2008). Tequila, mezcal y cerveza: De México para el mundo. *Agric. Soc. Desarro.*, *5*(2), 143–150.
10. Ramírez, J., (1995). Los magueyes, plantas de infinitos usos. *Biodiversitas.*, *3,* 1–7.
11. Ortiz-Basurto, R. I., Pourcelly, G., Doco, T., Williams, P., Dornier, M., & Belleville, M. P., (2008). Analysis of the main components of the aguamiel produced by the maguey-pulquero (*Agave mapisaga*) throughout the harvest period. *J. Agric. Food Chem.*, *56*(10), 3682–3687.
12. Muñiz-Márquez, D. B., Contreras, J. C., Rodríguez, R., Mussatto, S. I., Wong-Paz, J. E., Teixeira, J. A.M., & Aguilar, C. N., (2015). Influence of thermal effect on sugars composition of Mexican Agave syrup. *CyTA J. Food.*, *13*(4), 607–612.
13. Romero-López, M. R., Osorio-Díaz, P., Flores-Morales, A., Robledo, N., & Mora-Escobedo, R., (2015). Chemical composition, antioxidant capacity and prebiotic effect of aguamiel (*Agave atrovirens*) during *in vitro* fermentation. *Rev. Mex. Ing. Quim.*, *14*(2), 281–292.
14. Silos-Espino, G., González-Cortés, N., Carrillo-López, A., Guevaralara, F., Valverde-González, M. E., & Paredes-López, O., (2007). Chemical composition and in vitro propagation of *Agave salmiana* 'Gentry.' *J. Hortic. Sci. Biotechnol.*, *82*(3), 355–359.
15. Leal-Díaz, A. M., Santos-Zea, L., Martínez-Escobedo, H. C., Guajardo-Flores, D., Gutiérrez-Uribe, J. A., & Serna-Saldivar, S. O., (2015). Effect of *Agave americana* and *Agave salmiana* ripeness on saponin content from aguamiel (agave sap). *J. Agric. Food Chem.*, *63*(15), 3924–3930.
16. Santos-Zea, L., Fajardo-Ramírez, O. R., Romo-López, I., & Gutiérrez-Uribe, J. A., (2016). Fast centrifugal partition chromatography fractionation of concentrated agave (*Agave salmiana*) sap to obtain saponins with apoptotic effect on colon cancer cells. *Plant Foods Hum. Nutr. (N.Y., NY, U.S.).*, *71*(1), 57–63.
17. Yen, C. H., Kuo, Y. W., Tseng, Y. H., Lee, M. C., & Chen, H. L., (2011). Beneficial effects of fructo-oligosaccharides supplementation on fecal bifidobacteria and index

of peroxidation status in constipated nursing-home residents -A placebo-controlled, diet-controlled trial. *Nutrition (N.Y., NY, U.S.).*, *27*(3), 323–328.
18. Hidaka, H., Eida, T., Takizawa, T., Tokunaga, T., & Tashiro, Y., (1986). Effects of fructooligosaccharides on intestinal flora and human health. *Bifidobact Microflora.*, *5*(1), 37–50.
19. Phuwamongkolwiwat, P., Suzuki, T., Hira, T., & Hara, H., (2014). Fructooligosaccharide augments benefits of quercetin-3-O-β-glucoside on insulin sensitivity and plasma total cholesterol with promotion of flavonoid absorption in sucrose-fed rats. *Eur. J. Nutr.*, *53*(2), 457–468.
20. Vitali, B., Ndagijimana, M., Maccaferri, S., Biagi, E., Guerzoni, M., & Brigidi, P., (2012). An in vitro evaluation of the effect of probiotics and prebiotics on the metabolic profile of human microbiota. *Anaerobe.*, *18*(4), 386–391.
21. Van den Heuvel, E. G., Muijs, T., Brouns, F., & Hendriks, H. F., (2009). Short-chain fructo-oligosaccharides improve magnesium absorption in adolescent girls with a low calcium intake. *Nutr. Res. (N.Y., NY, U.S.)*, *29*(4), 229–237.
22. Castro-Rodríguez, D., Hernández-Sánchez, H., & Fernández, J. Y., (2015). Probiotic properties of *Leuconostoc mesenteroides* isolated from aguamiel of *Agave salmiana. Probiotics Antimicrob. Proteins*, *7*(2), 107–117.
23. Castro-Rodríguez, D. C., (2013). Estudio de la capacidad probiótica de las cepas de bacteria lácticas aisladas de *Agave salmiana* sp. MSc Dissertation, Instituto Politécnico Nacional. México D.F.
24. Estrada-Godina, A. R., Cruz-Guerrero, A. E., Lappe, P., Ulloa, M., García-Garibay, M., & Gómez-Ruiz, L., (2001). Isolation and identification of killer yeasts from Agave sap (aguamiel) and pulque. *World J. Microbiol. Biotechnol.*, *17*(6), 557–560.
25. Correa-Ascencio, M., Robertson, I. G., Cabrera-Cortés, O., Cabrera-Castro, R., & Evershed, R. P., (2014). Pulque production from fermented agavc sap as a dietary supplement in Prehispanic Mesoamerica. *Proc. Natl. Acad. Sci. USA.*, *111*(39), 14223–14228.
26. Osorio-Diaz, P., Sanchez-Pardo, M. E., & Bello-Perez, L. A., (2014). Mexican bakery products. *Bakery Products Science and Technology* (2nd edn., pp. 723–734). John Wiley & Sons, Ltd, Chichester, UK.
27. Rios, A., (2017). Pan de Pulque: el regalo de Saltillo a la gastronomía. Periodico vanguardia [Online]. http://www.vanguardia.com.mx/articulo/pan-de-pulque-el-regalo-de-saltillo-la-gastronomia. (Accessed 21 July 2017).
28. Santos-Zea, L., Leal-Diaz, A. M., Cortes-Ceballos, E., & Alejandra Gutierrez-Uribe, J., (2012). Agave (*Agave* spp.) and its traditional products as a source of bioactive compounds. *Curr. Bioact. Compd.*, *8*(3), 218–223.
29. Gutierrez-Uribe, J. A., & Serna-Saldivar, S., (2013). Agave syrup extract having anticancer activity. U.S. Patent No 8,470,858, 25 June 2013.
30. Silos-Espino, H., Tovar-Robles, C. L., González-Cortés, N., Méndez-Gallegos, S. J., & Rossel-Kipping, D., (2010). IX simposium-taller nacional y ii internacional de producción del nopal y maguey. estudio integral del maguey (*Agave salmiana*): Propagacion y valor nutricional. Escobedo (Nuevo Leon), México, November 12 and 13.

31. Tovar, L. R., Olivos, M., & Gutierrez, M. E., (2008). Pulque, an alcoholic drink from rural Mexico, contains phytase. Its in vitro effects on corn tortilla. *Plant Foods Hum. Nutr. (NY, U.S.)*, *63*(4), 189.
32. Valadez-blanco, R., Bravo-Villa, G., Santos-Sánchez, N. F., Velasco-Almendarez, S. I., & Montville, T. J., (2012). The artisanal production of pulque, a traditional beverage of the Mexican highlands. *Probiotics Antimicrob. Proteins, 4*, 140–144.

CHAPTER 6

NUTRACEUTICALS AS ADJUVANT TREATMENTS FOR CANCER AND NEURODEGENERATIVE DISEASES

MELISSA FLORES-GARCÍA,[1] MAYELA GOVEA-SALAS,[1] RODOLFO RAMOS-GONZÁLEZ,[2] SONIA YESENIA SILVA-BELMARES,[3] RAÚL RODRÍGUEZ-HERRERA,[3] ELDA PATRICIA SEGURA-CENICEROS,[1] ALMA ROSA PAREDES-RAMÍREZ,[4] DALIA VÁSQUEZ-BAHENA,[5] and ANNA ILINÁ[1]

[1] *Autonomous University of Coahuila, Nanobioscience Group, School of Chemistry, Blvd. V. Carranza and José Cárdenas Valdés s/n Col. Republic East, ZIP 25280, Saltillo, Coahuila, Mexico, Tel.: +52-844-416-92-13, E-mails: annailina@uadec.edu.mx; anna_ilina@hotmail.com*

[2] *Autonomous University of Coahuila. Blvd. V. Carranza and José Cárdenas Valdés s/n Col. Republic East, ZIP 25280, Saltillo, Coahuila, Mexico*

[3] *Autonomous University of Coahuila, Food Research Department, School of Chemistry. Blvd. V. Carranza and José Cárdenas Valdés s/n Col. Republic East, ZIP 25280, Saltillo, Coahuila, Mexico*

[4] *Autonomous University of Coahuila, Laboratory Animal Center, School of Medicine. Francisco Murguía South No. 205. ZIP 25000, Saltillo, Coahuila, Mexico*

[5] *Healthcare Business and Computer Technology S.A. de C.V. Tlaxcala No. 146/705, Col. Roma South ZIP. 06760, Cuauhtémoc, Mexico City, Mexico*

ABSTRACT

In the last years, nutraceutical products consumption has increased, mostly because of its natural origin, affordable prices, and benefits generated by its ingestion. Overall, nutraceuticals are formulated with different ingredients such as amino acids, proteins, vitamins, minerals, polyphenols, probiotics, and other bioactive compounds. These kinds of compounds can help the human body to maintain a physiological balance and therefore, health. Recently, nutraceuticals consumption on a regular basis has proven being a revolution compared to conventional pharmaceuticals basically, because of its curative and preventive potential against different pathologies. Research on nutraceutical products and its benefits are often discussed; however, there is some outstanding information that is certainly significant to assess regarding its performance against different kind of cancer pathologies and neurodegenerative diseases by its bioactive ingredients, which confers them certain helpful capacities. Accordingly, this manuscript focuses on the potential activity of these bioactive compounds with nutraceutical value as adjuvant treatments for important and outstanding pathologies. Some of the most important include breast cancer, colorectal cancer, lung cancer, Alzheimer and Parkinson's disease, among others, that are significant to evaluate and undoubtedly will contribute to the improvement of nutraceuticals research and development.

6.1 INTRODUCTION

Currently, the increased consumption of natural products is in a critical phase. Human concern for disease prevention rather than healing has been a crucial factor, which is necessary to achieve success with nutraceuticals products. It should be noted that human health is closely related to diet and therefore with the participation of biologically active dietary products; some of these products with nutraceutical potential, contribute to homeostasis.

There are many definitions for nutraceuticals; commonly, it is used the description reported by DeFelice [1] who mentioned that "nutraceutical" is made by combining "nutrition" and "pharmaceutical" [2], this because nutraceuticals are foods or food extracts administered through pharmaceutical matrices which possess biological activity in the body. This term was coined by the Foundation for Innovation in Medicine in 1989 to describe the area of biomedical research of some natural compounds that began to flourish during the period [1].

Frequently, the nutraceutical term is associated or confused with the concepts of dietary supplement, functional food, and medical food, among others (Table 6.1). It is common that these terminologies are related because of the description is given by the authors who define them according to regulations that exist in each country for these kinds of products [3, 4]. It is noteworthy that some authors describe and consider a nutraceutical as food supplements since both products are defined as compounds isolated from a natural source; such as food or plants that are eaten as part of a healthy conventional diet [5, 6]. However, others report that the most significant difference is that nutraceuticals are not held in food matrix as functional foods, but rather in a pharmaceutical form such as tablets, powders, pills, among others, even though these might depend in the context of the definition of some authors [7].

TABLE 6.1 Description of Some Terms Associated with Nutraceuticals

Term	Description
Functional foods	Any food that consumed has a nutritional value and positive impact on health stability. Found in a food matrix, but may be altered by removing, adding or replacing some of its components that make them better than the originals.
Dietary or food supplements	Products consumed as part of a healthy regime, not found exclusively in a food matrix. The purpose of this kind of product is not based on improving health or pathologies compared to nutraceuticals.
Nutraceuticals	Products consumed as part of a healthy diet but its incorporation is not limited to a food matrix. Found in pharmaceutical forms such as pills, powders, tablets or others. They can prevent or cure diseases.

6.2 NUTRACEUTICALS CLASSIFICATION

Most nutraceuticals are composed of bioactive compounds (Table 6.2), which are obtained from diverse parts of plants such as fruits, vegetables, and even microorganisms [8, 9]. Furthermore, it has been shown that bioactive compounds can perform different biological activities (Table

6.2). Classification of nutraceuticals is difficult because of a large number of ingredients that this kind of products has [10]. However, Singh and Sinha [11] stated that nutraceuticals can be divided into two groups: the "traditional" including all-natural substances obtained from foods, without being altered in any way whether biochemistry or biotechnologically; and the "non-traditional" which include those modified with bioengineering methodologies.

TABLE 6.2 Common Bioactive Compounds in Nutraceuticals and Their Biological Activities

Bioactive compounds	Examples of biological activities
Amino acids Proteins Vitamins Minerals Carbohydrates Polyphenols Fatty acids Prebiotics Probiotics Dietary fiber Carotenoids	Antioxidant Anti-inflammatory Anti-allergic Immunomodulation Anti-proliferative Cancer prevention Lipids control (Lowering lipids blood levels) Cardioprotective Diabetes control (Lowering glycemic blood levels) Neurodegenerative diseases prevention

On the other hand, Lokesh et al. [12] indicated that nutraceuticals can be grouped into four different categories, the first group includes some dietary nutrients (carbohydrates, amino acids, vitamins or minerals); the second is dietary supplements containing a mixture of food nutrients in the first group. The third category includes non-toxic secondary metabolites obtained as result of extractions of healthy plants or herbs whose beneficial properties have been already reported. Finally, the fourth group includes herbal extracts or extracts of traditional medicines whose beneficial biological activity has also been tested and reported. Generally, the nutraceuticals classification can be summarized in Table 6.3.

6.3 NUTRACEUTICALS DEVELOPMENT

Nutraceuticals are consumed because they offer concentrated bioactive compounds, which can act in the human body and improve its health,

TABLE 6.3 Overall Classification of Nutraceuticals

Classification system	Description
Pandey et al. [13]	• Potential: Viable products for its use as a nutraceutical but does not have enough clinical data to prove a health benefit. • Established: Nutraceuticals products that are already reported to have health benefits.
Singh and Sinha [11]	• Traditional: Nutrients obtained from foods without being altered in any way. • Non-traditional: Nutrients modified by bioengineering procedures.
Chauhan et al. [9]	• Natural: Extracted from natural sources such as plants, foods or microorganisms. • Chemical: Synthesized from chemical reactions.
Lokesh et al. [12]	• First group: Dietary nutrients. • Second group: Dietary supplements (mix of first group ingredients). • Third group: Non-toxic secondary metabolites obtained from plant extracts. • Fourth group: Herbal extracts and/or traditional medicine extracts.

by either treating or preventing some diseases [4, 6, 7, 9]. Production of nutraceuticals involves the use of natural bioactive compounds obtained from different sources including: plants, foods, and microorganisms; some of the technologies used for the extraction of these compounds are listed in Table 6.4 [14, 15].

Biologically active substances can be obtained by diverse methods of extraction thus different methodologies can be employed to isolate, concentrate, and purify these compounds. Accordingly, the method used depends on the product and degree of desired purity to be obtained. However, in

TABLE 6.4 Methodologies Used for Bioactive Compound Extractions in Different Sources

Extraction procedures for bioactive compounds
Solvent or enzyme-assisted extraction
Pressurized liquid extraction
Subcritical fluid extraction
Supercritical extraction
Microwave-assisted extraction
Ultrasonic-assisted extraction
Pulsed-electric field extraction
Instant controlled pressure drop-assisted extraction

most cases, purification methods used are based on the employment of some chromatography technique [16, 17].

Considering that, bioactive compounds include a broad diversity of molecules; each one of them specialized in developing a biological activity in the human body, it is necessary to prove its biological influence and toxicological security for their consumption as part of the diet [18]. Every bioactive compound used as nutraceutical constituents are not dangerous according to their natural extraction sources since people usually ingest them conventionally in the form of food or are found in harmless sources; plus, the regulation systems and laws for their commercialization are not strict enough as for pharmaceutical products [6, 9].

Some of the common methodologies to guarantee its safety include *in vitro* analysis with different cell lines or biological models such as *Artemia salina*, *Daphnia magna* and *Danio rerio* [19. 20]. Occasionally, strategies include the not so frequent but more advantageous *in vivo* examination using different biological models such as rabbits, rats, but mostly, mice [18, 21]. Subsequently to these strategies, it has been reported that some researchers have implemented *in vivo* human models analysis known as "nutritional interventions" to study toxicology, biological effects, pharmacokinetics, and etcetera. However, this kind of studies are not frequent because of the difficulties that they represent such as time-consuming analysis, multiple steps methodologies which are expensive, but firstly, human security and ethical issues [18, 21, 22].

A general scheme of the development of nutraceuticals can be found in Figure 6.1.

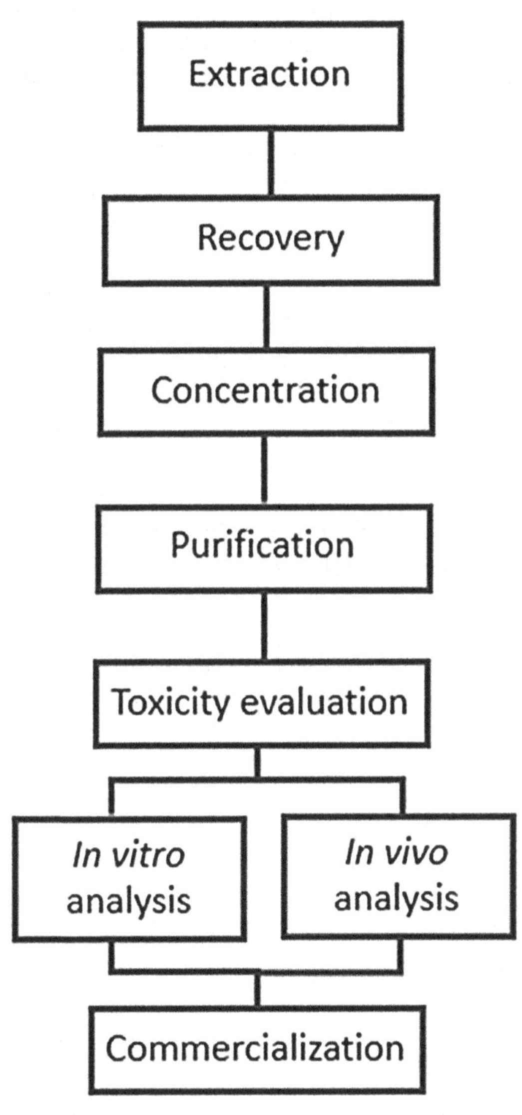

FIGURE 6.1 General scheme of the development of nutraceuticals.

6.4 BIOACTIVE COMPOUNDS IN NUTRACEUTICALS

As it was mentioned before, nutraceutical products are composed of different kinds of bioactive compounds with the nutritional importance that can be found in diverse sources in nature. Some bioactive compounds with nutraceutical value are shown in Table 6.5 with different examples and their sources can be found as well.

Most bioactive compounds found in nutraceutical products possess multiple therapeutic benefits and are related to a general physiological profit. Nowadays, the main target for these compounds are outstanding pathologies such as cancer and/or neurodegenerative diseases where conventional pharmacological systems are not always effective or can generate secondary effects that have repercussions in health. Strong evidence in the field of health has been reported about how bioactive compounds, nutraceuticals, functional foods, among others, can achieve certain positive biological activities as a result of its consumption [47].

6.5 NUTRACEUTICALS FOR CANCER AND NEURODEGENERATIVE DISEASES

According to the World Health Organization (WHO) [48], cancer is the principal cause of death in the world, at least in 2015, the number of deaths occurred by this pathology was 8.8 million. Some of the most important types of cancer are:

1 lung cancer;
2 hepatic cancer;
3 colorectal cancer;
4 gastric cancer; and
5 breast cancer.

6.5.1 LUNG CANCER

This kind of cancer is the first cause of deaths by cancer worldwide; it can be classified either as a small cell lung cancer (SCLC) or as a non-small cell lung cancer (NSCLC). The primary treatment scheme for most patients

TABLE 6.5 Examples of Bioactive Compounds with Nutraceutical Value and Their Sources

Bioactive compound	Example	Source	Reference
Amino acids	Glycine Aspartic acid Glutamic acid Tryphtophan	Pistachios Maize Codfish Pilchard	Anbazahan et al. [23] Je et al. [24] Vijay et al. [25]
Proteins	Caseine Serum albumin Whey protein	Milk Cheese Butter	Valiño et al. [26] Kimpel and Schmitt [27] Diarrassouba et al. [28]
Vitamins	Vitamin A Vitamin C Vitamin D Vitamin E	Carrots Lemons Salmon Spinach	Nieves [29] Houston [30]
Minerals	Potassium Calcium Magnesium Zinc Iodine	Bananas Eggs Nuts Chicken Cranberries	Houston [30] Anbazahan et al. [23]
Carbohydrates	Starch Cellulose Fructose Maltose	Potatoes Lettuce Cauliflower Bread	Webster and Lehrke [31] Pallela [32]
Polyphenols	Gallic acid Coumaric acid Sinapic acid Resveratrol Caffeic acid	Pomegranate Tomatoes Blueberries Strawberry Coffee	Bahadoran et al. [33] López-Gutiérrez et al. [34] Tomé-Carneiro and Visioli [35]
Fatty acids	Omega–3 fatty acids: Alpha-linoleic Eicosapentaenoic acid Docosapentaenoic acid Docosahexaenoic acid	Anchovies Salmon Shrimp Egg yolk Canola oil	Alanazi [36] Hamilton et al. [37]
Prebiotics	Oligosaccharides: Xylooligosaccharides	Cabbages Onions	Ganguly [38] Samanta et al. [39] Wichienchot et al. [40]
Probiotics	Lactic acid bacteria: *Lactococcus,* *Streptococcus* *Bifidobacterium*	Yogurt Pickles Microalgae Pickled cucumbers	Behnsen et al. [41] Begum et al. [42]

TABLE 6.5 *(Continued)*

Bioactive compound	Example	Source	Reference
Dietary fiber	Inulin Lignin Chitins Pectins Beta-glucans	Garlic Artichoke Asparagus Apple Lemons	Trung and Stevens [43]
Carotenoids	Lycopene Lutein Zeaxanthin Fucoxanthin Canthaxanthin Astaxanthin	Tomatoes Watermelon Carrots Wakame algae Cabbages Microalgae	Tanaka et al. [44] Berrow et al. [45] Gonzalez-Sarrias et al. [46]

who have SCLC involves chemotherapy and, in particular cases, radiation therapy. On the other hand, NSCLC patients undergo surgery as a first medical action but in some cases, patients also receive chemotherapy and/ or radiation [49]. Although surgery, chemotherapy, and radiation therapy (alone or combined between them) are a strong but practical intervention scheme for lung cancer, medication such as erlotinib and afatinib can positively improve the results of the treatments. However, these medications can develop secondary adverse effects from simplc rash, diarrhea, conjunctivitis, headaches, among others, to a hepatic or kidney failure, intestinal bleeding, etc. [50, 51].

Regarding bioactive compounds that have nutraceutical value, there is plenty of information which mentions that the anti-oxidative activity and other molecular pathways in the body help to suppress angiogenesis and/ or carcinogenesis of the tumor cells that produce lung cancer and even can help with adverse side effects of conventional treatments [52]. Recently, Ku et al. [53] indicated that maclurin, an organic compound extracted from the *Moruslaba* (white mulberry) and *Garcinia mangostana*, had the antioxidative capacity with anti-metastatic effects of human NSCLC cells on an *in vitro* level. On the other hand, Wright et al. [54] studied the effects of a daily antioxidant intake such as carotenoids, flavonoids, and vitamin E; in this investigation, it was remarkable that lung cancer risk in subjects was decreased by consumption of these bioactive compounds.

To conclude, reports in the specific nutraceutical field, such as the investigation by Frese et al. [55] showed the sensitization of cancer lung cells to apoptosis by the influence of the Oleandrin. Also, other reports indicate the effect of Berberine, a nutraceutical found in different parts of some plants such as *Berberis vulgaris*, *B. aristata*, *B. aquifolium*, among others, inducing growth inhibition and apoptosis of NSCLC cells on an *in vivo* level [56].

6.5.2 HEPATIC CANCER

There are two different kinds of hepatic cancer: primary and secondary; the first one is referred to cancer that originates in the liver, while the secondary is the one where cancer originates in another organ and then it spreads to the liver [57]. Hepatocellular carcinoma (HCC) is the most common form of primary hepatic cancer and the second cause of death worldwide [58] and the risk factors include infection with hepatitis B and hepatitis C viruses (HBV and HCV). However, there is also influences like host genetic factors and others associated with the person lifestyle such as heavy alcohol intake, obesity, smoking, and less frequently, consumption of food contaminated with aflatoxins [59].

Main causes of hepatic cancer can be prevented through simple public health services, which include appropriate vaccination against HBV and HCV, sanitary practices and healthy life choices. Nevertheless, once HCC is present, it mainly depends on the advanced state in which it is to be treated. First stages of the HCC include curative treatments involving ablation, resection or liver transplants, intermediate to advanced stages may also involve palliative treatments such as chemoembolization and some immune-based pharmaceuticals involvement [60, 61].

Since HCC is essentially incurable cancer, much research has been made to develop adjuvant therapies that help the patient to overcome secondary side effects from conventional treatments and, ideally, aid the host to improve their health. One of these strategies is the use of nutraceuticals and/or bioactive compounds. In this context, investigations by Sethi et al. [62] using the bioactive compound garcinol extracted from *Garcinia indica* fruit assisted to suppress growth of human HCC on *in vitro* and *in vivo* studies. On the other hand, use of nutraceuticals such as quercetin, perillyl alcohol, fisetin, curcumin, epigallocatechin gallate, eriodictyol, and naringenin can inhibit the HCC angiogenesis and metastasis by their anticytokine mechanism of action [63].

6.5.3 COLORECTAL CANCER

This type of malignancy is one of the most common cancers and is the third cause of death worldwide. According to WHO [64], if no action is taken, the incidence of colorectal cancer is expected to rise by 60% in 2030. Causes of colorectal cancer are multifactorial, however, most of them are related to dietary factors. In this sense, preventive measures for colorectal cancer include healthy life choices such as doing exercise and maintain healthy diets, avoid smoking and alcohol consumption may be the difference.

Typically, treatment depends on the pathology stage and tumor localization. Surgery to remove it, is the first step for early stages; a complete colostomy is not an often procedure, but when done, it is usually a temporary form of treatment. For its part, advanced stages of cancer include treatments that involve radiotherapy, chemotherapy, immunotherapy, and targeted therapy [49]. It is also noteworthy that, even though there is a high survival rate for early stages of treated cancer, most survivors report that their physical quality of life is lower than that of general healthy population [65].

Naturally, as a cancer type, treatment also includes synthetic anticancer drugs. However, this kind of pharmaceutical choices have secondary adverse effects and off-target activities [49, 66]. Thus, there is an opportunity area for the nutraceutical industry. Reports by Kuppusamy et al. [66] indicate that several nutraceutical products such as dietary supplements, plant secondary metabolites, medicinal herbs and even microorganisms can contribute to health improvement in the colorectal disease. Likewise, bioactive compounds with nutraceutical value such as curcumin, grape seed extract, and rhizochalin, to name a few, were reported to effectively control progression of colorectal cancer [67–69]; while oleuropein is reported to be a good nutraceutical for prevention of the same pathology [70].

6.5.4 GASTRIC CANCER

This type of cancer produces the death of 774 thousand persons annually, making it the fourth cause of death worldwide according to WHO [48]. Approximately, 90% of gastric cancers are adenocarcinomas while the rest are other types of cancer that arise from the stomach and muscles surrounding the mucosa of the same organ [71]. The risk factors of this cancer include environmental and genetic factors making it a sporadic disease that can be initiated by longstanding diseases such as gastritis, which is caused by *Helicobacter pylori* infection. Nevertheless, lifestyle habits like smoking and some dietary influences such as intake of salty and smoked foods, and low consumption of fruits and vegetables may take a part in the gastric cancer development.

Regarding nutraceutical products role, studies have shown that some dietary factors involving intake of antioxidants and vitamins may be effective to prevent gastric cancer. Qiao et al. [72] indicate that selenium, vitamin E, and β-carotene consumption played a significance role decreasing mortality of gastric cancer patients. In the same way, Serafini et al. [73] showed that a high intake of antioxidant compounds obtained from different plants is clearly associated with a reduction of gastric cancer.

More recently, Ullah et al. [74] reported that nutraceuticals such as resveratrol, lycopene, curcumin, genistein, γ-aminobutyric acid (GABA), and epigallocatechin-3-gallate (EGCG) interfere with progression of carcinogenesis in gastrointestinal cancers by different action mechanisms.

6.5.5 BREAST CANCER

This type of pathology is the most frequently diagnosed cancer accounting for the 25% of all cancer cases. According to WHO [48], is the fifth cause of death worldwide with an estimate of 571 thousand people dead by the end of 2015 causing 15% of all cancer deaths among females.

Respectively, treatment for breast cancer usually involves surgical interventions: breast-conserving surgery (BCS) or mastectomy. Of all cases, 59% of females that have early stages of breast cancer experience a BCS, 36% have a mastectomy, 4% is treated with radiotherapy and/or chemotherapy, while the remaining 1%, does not receive any kind of treatment [75]. On the other hand, patients with late stages of the disease receive chemotherapy with or without surgical interventions.

Despite the relative high survival rate of early stages of breast cancer that involves radiation therapy and surgical intervention, this practice develops long-term side effects including numbness and chronic pain [76, 77]. In addition, patients who receive chemotherapy can develop infertility or premature menopause. Other adverse side effects generated by breast cancer treatment include osteoporosis, myalgia, arthralgia, and chronic fatigue, to name a few [78, 79].

In this way, development of new products that reduce toxicity and collateral side effects of conventional treatments for breast cancer is imperative. Novel studies indicate that using bioactive natural compound thymoquinone, which is isolated from *Nigella sativa*, as a nutraceutical product, achieves anticancer activity against a breast cancer cell line *in vitro* [80]. On the other hand, reports by Mock et al. [81] show that curcumin and curcumin analogs have antiproliferative, apoptotic, and anti-tumor activity against breast cancer cells, and that, in fact, mixing them with other therapeutic anticancer agents creates a synergistic effect against this disease. Lastly, other studies report that a broad range of nutraceuticals obtained from different plants and foods like garlic and soybean possess activity against breast cancer [82].

Alternatively, bioactive compounds in nutraceuticals as adjuvants or treatments *per se,* can improve health in pathologies that affect neurological systems such as Alzheimer's disease (AD) and Parkinson's disease (PD).

6.5.6 ALZHEIMER'S DISEASE

It is a degenerative and progressive brain pathology produced because neurons in some parts of the brain are no longer functional. The AD is the most common cause of dementia and is characterized by a deterioration in memory, language, and other cognitive abilities that affects developing of ordinary day-to-day activities and that, eventually, can affect basic body functions such as walk or swallow. Thus, the AD is ultimately known as a fatal disease [83].

The AD is a multifactorial pathology, however, most experts in the field agree that, besides some genetic specific mutations, influences like age, body weight, type 2 diabetes, metabolic syndrome, smoking, among others, may play a role in the AD development [84].

According to the reports by the Alzheimer's Association [83], there are six pharmacological treatments approved to temporally improve AD symptoms by increasing the number of neurotransmitters in the brain. However, this pharmaceutical scheme does not slow nor stop AD progression. It is noteworthy that the development of new pharmacological treatments is extremely difficult because of high production costs and the long time required to be approved by health organisms accordingly to the country in which is developed. Also, there is a non-pharmacological treatment that includes exercise, music listening, and reminiscence therapy, to name a few, but this kind of therapy is focused on cognitive function maintenance.

6.5.7 PARKINSON'S DISEASE

This is the second most common neurodegenerative disease and is caused by degeneration of neurons in the brain with yet unknown causes; this chronic and progressive pathology is characterized by resting tremor, rigidity, slow movements, postural, and autonomic instability [85]. Although PD is an incurable illness, some pharmacological treatments such as levodopa, dopamine agonists, β-blockers, among others, may improve life quality for many years [86].

For both outstanding neurodegenerative diseases, much research has been done to develop new complementary and alternative treatments and, even though there is some skepticism in the capacity of nutraceutical compounds to help homeostasis, other reports indicate that there is a remarkable response, especially in disease prevention. In this sense, nutraceutical products may be an important key in search of early therapeutic and preventive approaches. In the nutraceutical field, reports using crocin, lycopene, carnosic acid, rosmarinic acid, and resveratrol have shown positive results for AD [87–90], while anethole, thymoquinone, carnosol, and kaempferol [91–94], to name a few, are nutraceuticals beneficial to reduce PD risk.

6.6 ACTION MECHANISM OF NUTRACEUTICALS FOR HEALTH BENEFITS

Nowadays, much attention has been paid to the relationship between nutrition and health so, many natural products have been developed to achieve a healthy diet. In this context, marketing of nutraceutical products has proven to be an excellent strategy because of the natural origin of its bioactive constituents and ability to cure and prevent disease. However, most nutraceuticals producers do not pay attention to the action mechanism of bioactive compounds; therefore, it is unknown how they achieve their biological activity, physiological balance in the body, and consequently, homeostasis. Some of the reported action mechanisms for nutraceutical products are listed below:

1. Enzymatic reactions (including activities as substrates, cofactors, and inhibitors).
2. Enhance absorption and/or stability of essential components (including nutrients or pharmaceutical molecules).
3. Enhancers of beneficial bacteria and/or inhibitors of deleterious bacteria.
4. Scavengers of reactive or toxic chemicals.

6.6.1 ENZYMATIC REACTIONS

One of the principal mechanisms that some nutraceutical products have is their capacity to act as components of biochemical reactions. Some of the bioactive compounds in the formulation of nutraceuticals seem to be involved in enzymatic reactions either acting as substrates, cofactors or inhibitors.

Enzymes are molecules that catalyze biochemical reactions. The reaction derives from the association of chemically reactive groups within active sites of enzyme and substrates (target molecule), which will trigger reaction steps required to convert the substrate into a reaction product [95]. Some enzymes do not require an additional component to develop their activity while others are associated with non-protein molecules known as cofactors. Furthermore, conversion of substrate into a reaction product can be inhibited by the presence and juxtaposition of inhibitors into enzyme active site. The enzymatic activity is generally described in Figure 6.2.

It is worth mentioning that activity (substrate, cofactor or inhibitor) strongly depends on enzyme type. However, FitzGerald and Meisel [96] stated that casokinins and lactokinins, a group of bioactive peptides, could inhibit angiotensin-I-converting enzyme (ACE), which is responsible for regulation of peripheral blood pressure. Also, it is reported that luteolin flavonoid inhibits metalloproteases involved on articular cartilage degradation and could be used as a nutraceutical for arthritis treatment [97].

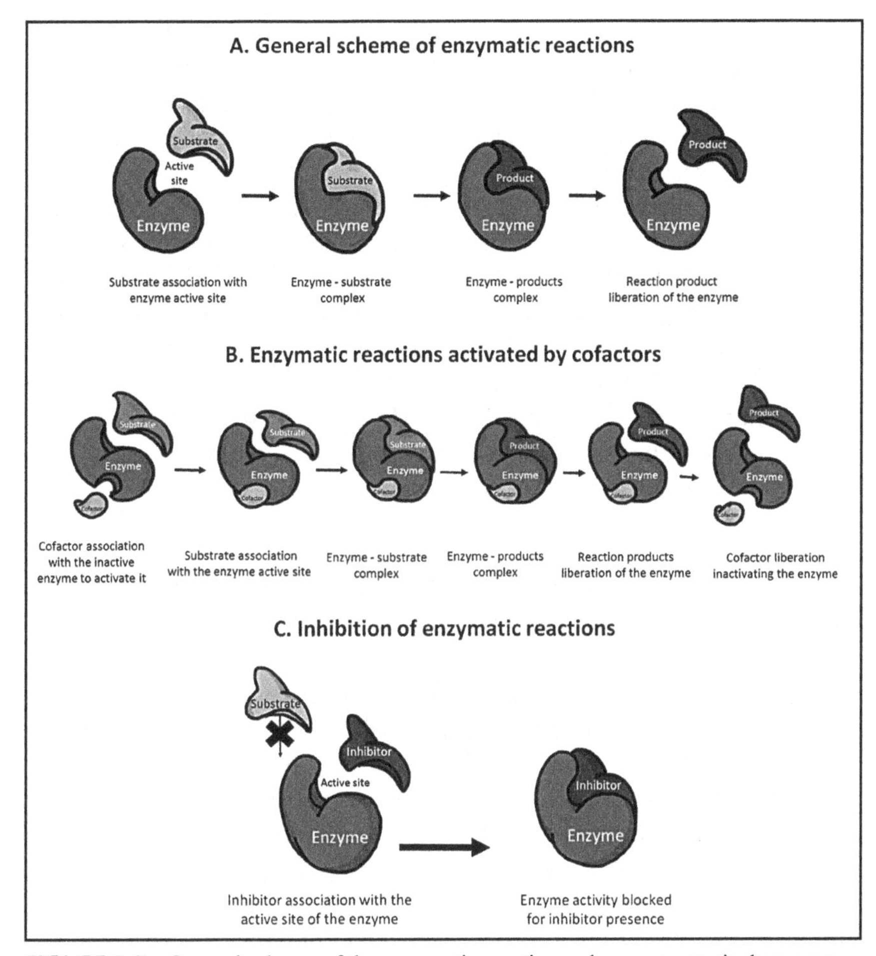

FIGURE 6.2 General scheme of the enzymatic reactions where nutraceuticals can act either as substrates or inhibitors.

Some of enzymes contained in nutraceuticals are related to dietary enzymes that help gastrointestinal system to absorb and digest the food. Faisal and Varma [98] mentioned that lot of digestive enzymes such as papain, bromelain, pancreatin, and trypsin (which come from different sources such as animals, fruits, and plants) are used in nutraceutical products in order to beneficiate digestive system. In another way, Menon and Lele [99] reported that many bioactive compounds such as enzyme inhibitors and proteolytic enzymes (pepsin, calpains, and collagenases) obtained from fishery sources and seafood wastes are used in nutraceuticals development. Similarly, Díaz-López and García-Carreño [100] stated that enzymes from fish and shellfish residues are being used in food products and their consumption is increasing mostly because of their curative potential against diseases.

In contrast, there are also reports that indicate that bioactive compounds found in nutraceuticals can act in a transcriptional level affecting transcription factors, which lead to expression or depression of enzymes. For example, a specific interaction between a flavonoid and transcriptional factor NF-E2-related factor 2 (Nrf2) is been reported by Shih et al. [101], this mechanism activates transcription of antioxidant enzymes that help the system to decrease oxidative stress. However, these kinds of activities are hard to report because of lack of investigation of nutraceuticals and its bioactive compounds effects at biomolecular level.

6.6.2 ENHANCERS OF ABSORPTION AND/OR STABILITY OF ESSENTIAL COMPONENTS

Consumption of nutraceutical products offers concentrated bioactive compounds that can improve human health by treating or preventing diseases. Yet, nutraceutical ingestion on a daily basis by healthy people increases the consumption of some specific nutrients that will aid the conventional food regime assuring to meet organism necessities. Reports by Nijveldt et al. [102] state that intake of flavonoids as nutraceuticals exceed those of vitamin E and β-carotene, this result suggests that the impact of the nutraceutical products is increasing day by day. Besides, the nutraceutical products can also potentiate the effect of conventional pharmaceuticals as reported by Campolongo et al. [103], their study indicates that simultaneous administration of a nutraceutical and a pharmaceutical enhances the effect of the lipid normalization levels in patients with ischemic heart disease. Similarly, Pan et al. [104] recognize that some bioactive compounds can be used sensitizers to enhance the efficiency of other pharmaceuticals agents against cancer; however, the action mechanism still has to be elucidated to understand the approaches of this kind of bioactive compounds.

One of the main problems of nutraceuticals delivery systems for their consumption is the poor absorption they have in the digestive system which limits their bioavailability [104], for that reason, the ingestion of bioactive compounds that can be easily absorbed and can be stable in the organism remains as a primary objective for nutraceuticals researchers.

According to Davinelli et al. [105] the use of nutraceuticals as part of the scheme treatment for Alzheimer's disease can help to stabilize the genetic material (DNA) of the patients avoiding the epigenetic changes that cause the cognitive decline and neuronal dysfunction, this by inhibiting the process of DNA methylation and histone acetylation. Equally, studies have shown that patients with Parkinson's disease that ingest dietary fiber (considered bioactive compounds in nutraceuticals) reported to beneficially alter gut microbiota and intestinal permeability increasing nutrients absorption [106]. In other studies, it is reported that bioactive compounds found in garlic can interact in a cellular level inhibiting the dietary absorption of lipids helping the organism in lipid-metabolism disease cases [107].

6.6.3 ENHANCERS OF BENEFICIAL BACTERIA AND/OR INHIBITORS OF DELETERIOUS BACTERIA

In the present chapter, it is reported that some bioactive compounds are prebiotics and probiotics, which can be used as part of formulation in nutraceutical products. Prebiotics are non-digestible and selective fermented bioactive compounds often obtained from food sources; these products affect gut microbiota stimulating growth or activity of intestinal bacteria [108, 109]. To be considered a prebiotic, compound must have some characteristics such as resistance to gastric acidity and enzymes, susceptibility to be fermented by gut microbiota and ability to stimulate growth and/or activity of beneficial intestinal bacteria [110]. On the other hand, probiotics are defined as microorganisms, which once ingested in a certain way can exert health benefits beyond the ones obtained from essential nutrition [111]. Some of the characteristics that probiotics must have are nonpathogenic, nontoxic, resistance to gastric acidity and enzymes, adherence to gut epithelial tissues, temporary colonization of intestine, beneficial to host, production of antibacterial substances, and ability to inhibit pathogenic strains [109].

Firstly, prebiotics use in nutraceutical products is intended to work directly on intestinal microbiota; reports suggest that prebiotics beneficially nourish non-pathogenic bacterial strains acting as "growth factors," increasing their population to become the predominant genera in the intestine, which leads to improve colonization resistance against pathogenic bacterial strains and avoiding gastrointestinal infections and diseases [109, 112]. Reports by Laparra and Sanz [110] indicated that polyunsaturated fatty acids (PUFA) have beneficial effects on gut microbiota mainly because of their capacity to generate fermentation products that aid microorganisms to develop different kind of biological roles. Also, according to Akhtar et al. [113], many products extracted from pomegranate such as punicalagins, punicalins, gallic acid, and ellagic acid showed biological activity, increasing populations of certain non-pathogenic strains of bacteria so acting as prebiotics, and simultaneously decreasing population of pathogenic strains. The action mechanism for prebiotics is exemplified in Figure 6.3.

Probiotics action mechanisms are neither well reported nor understood. However, some specifications and reports are considered based on the documented analysis through *in vitro* and *in vivo* investigations, which allowed results extrapolation. Mainly, the action mechanism for

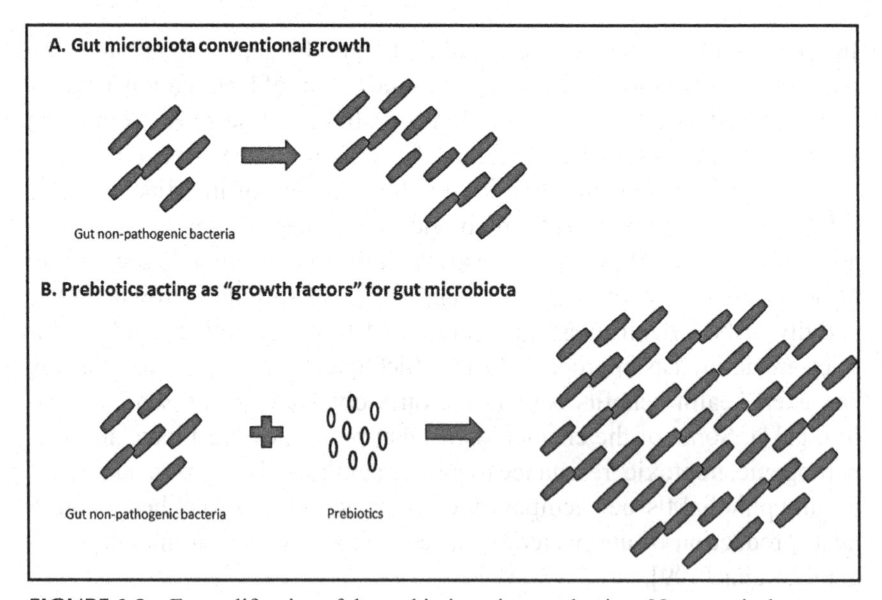

FIGURE 6.3 Exemplification of the prebiotic action mechanism. Nutraceutical prebiotics act as growth factors or substrates to enhance the gut.

probiotics is modulation of host microbiota; microbiological interventions through ingestion of probiotics are stabilizing or improving microbiota homeostasis by enhancing population of beneficial microorganisms and decreasing pathogenic microorganisms. Butel [114] reported three different mechanisms for probiotics action. The first mechanism is resistance to colonization, either preventing or limiting occupation of pathogen microorganisms. This mechanism can be established for pathogen bacterial inhibition because of the production of antibacterial compounds by probiotics, which lead to a decrease of environmental pH making difficult, the bacterial growth. Also, competition for nutrients can be displayed among different bacterial strains; in this case, probiotics can take over the pathogens since they are the predominant bacterial genera because of the probiotic action. The second mechanism is improving the barrier function of gut microbiota, where probiotics act at the cellular level involved in signaling pathways, leading to increase of intestine mucus layer, in addition to proteins production, which improves physiological barrier in the gastrointestinal system. Finally, the third mechanism is related to modulation of immune system since 70% of immune cells are located at the gut level.

The relationship established between non-digestible substrates known as prebiotics and beneficial microorganisms in gut microbiota known as probiotics, have led to positively modulate gastrointestinal health. In fact, some nutraceutical products and dietary supplements that use these kinds of bioactive compounds are reported to be formulated and ingested simultaneously to improve biological activity, this symbiotic formulation and inherent synergistic activity between prebiotics and probiotics is recognized as synbiotics products [111, 115].

6.6.4 SCAVENGERS OF REACTIVE OR TOXIC CHEMICALS

Some bioactive compounds have capacity to act as scavengers for either reactive or toxic molecules; this process can also be known as radical scavenging. Many bioactive compounds found in nutraceuticals, especially polyphenols, are reported to act as scavengers for noxious compounds in human body, known as free radicals. These groups of molecules are small chemical species that have an odd number of electrons and can act as ions for some reactions, they have a short lifetime and they do not need big quantities of activation energy to occur [116]. Generally, they can be oxygen free radicals or nitrogen free radicals, also known as reactive oxygen species (ROS) and reactive nitrogen species (RNS), respectively [117]; and includes diverse molecules (Table 6.6). The ROS and RNS species are normal products from cellular metabolism and both play beneficial and harmful roles [118]. However, it is important to describe that the harmful effect of species is called oxidative stress and/or nitrosative stress, respectively [117].

TABLE 6.6 Reactive Oxygen Species (ROS) and Reactive Nitrogen Species (RNS)

ROS		RNS	
Name	**Symbol**	**Name**	**Symbol**
Hydroxyl	$OH\cdot$	Nitrous oxide	$NO\cdot$
Superoxide	$O_2^{\cdot -}$	Peroxynitrite	$ONOO^-$
Nitric oxide	$NO\cdot$	Peroxynitrous acid	ONOOH
Thyl	$RS\cdot$	Nitroxyl anion	NO^-
Peroxyl	$RO_2\cdot$	Nitryl chloride	NO_2Cl
Lipid peroxyl	$LOO\cdot$	Dinitrogen trioxide	N_2O_3
Hydrogen peroxide	H2O2	Nitrous acid	HNO_2
Singlet oxygen	$^{-1}O_2$		
Hypochloric acid	HOCl		

Exposure to harmful effects to ROS and/or RNS species has led human organism to develop different defense mechanisms, among them, Valko et al. [118] reported that preventive and repair mechanisms, physical and antioxidant defense, are the main ones; for this work, antioxidant defense is the most important since they involve enzymatic and non-enzymatic antioxidants such as vitamins (vitamin C and E), carotenoids, flavonoids, and other polyphenols.

The principal way, in which bioactive compounds work against these free radical groups is by stabilizing molecules, being their target to oxidize and, afterwards, converting radical groups in less-reactive radicals [102]. Interestingly, Fraga et al. [119] reported that antioxidants like polyphenols perform their activity by a chain-breaking reaction of free radicals, this because of the phenolic hydroxyl groups that reduce free radicals through electrons donation; besides, polyphenol aromatic structures allow stabilization of free radicals by resonance. The general scheme for bioactive compounds reaction (e.g., polyphenols) with free radicals is showed in the follow equation:

$$Polyphenol\ (OH) + R' \rightarrow Polyphenol\ (O') + RH$$

where bioactive compounds contain hydroxyl groups (represented with OH), react with free radicals (represented as R'), in an electron donation leading to their stabilization.

There is a lot of evidence in which bioactive compounds contained in nutraceuticals act as antioxidants scavenging reactive or toxic chemicals such as ROS and RNS (Figure 6.4). For example, Prasad et al. [120]

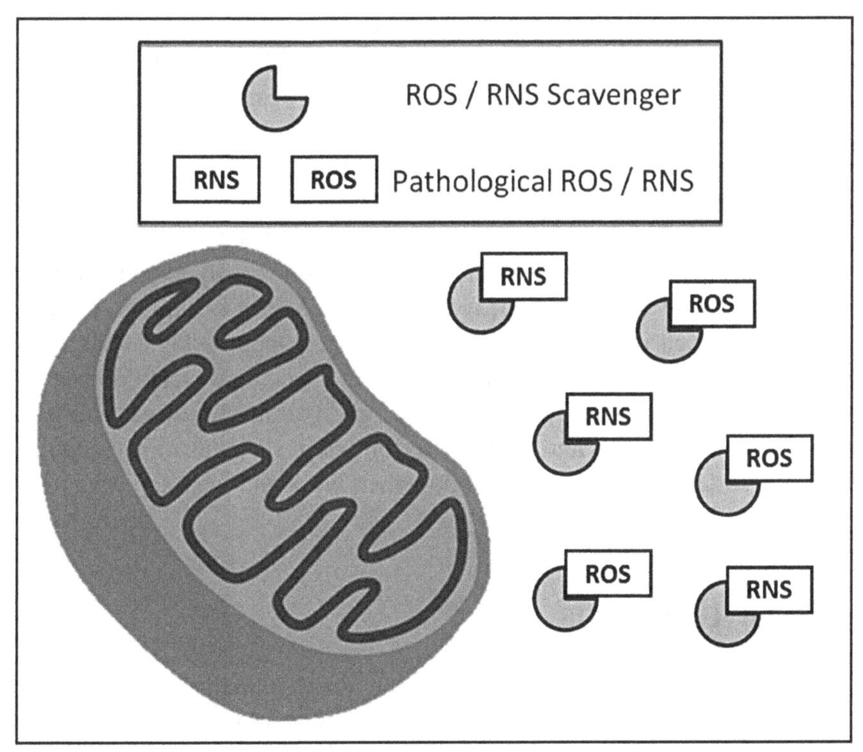

FIGURE 6.4 Exemplification of the ROS and RNS scavenger action mechanism of nutraceuticals.

mentioned that antioxidant bioactive compounds found in nutraceuticals are highly effective against harmful cancer effect of oxidative stress in the organism. In a very specific form, flavonoids, which are considered as bioactive compounds for nutraceuticals formulation, act as free radical scavengers for ROS and RNS forms [102, 121].

6.7 CONCLUSION

This work provided an up-to-date summary of the findings on nutraceutical compounds used as treatments or adjuvants for cancer and neurodegenerative diseases. The bioactive compounds encountered in nutraceuticals can provide a wide spectrum of biological activities triggered by different kinds of action mechanisms which lead to physiological balance preventing or treating malignancies, and therefore, contribute to homeostasis maintenance. The variety of nutraceuticals, as well as their action mechanism to treat and/or prevent, depends on the disease. In this analysis are shown a few action mechanisms that are related to the health benefits provided by nutraceuticals. However, it is possible to find others according to the compound involved; yet, models described in this review are some of the most important and reported.

Health benefits from bioactive compounds encountered in nutraceutical products are extremely diverse and are continuously expanded with new insights and scientific improvements, therefore, it is required to keep searching models in which these products develop their biologic activity in outstanding malignancies such as all kinds of cancer and incurable neurodegenerative pathologies like Alzheimer's and Parkinson's disease, as part of conventional studies and analysis performed in such nutritional supplements in order to fully understand how they work and what improvement can they offer.

ACKNOWLEDGMENTS

The authors would like to thank the Mexican Council of Science and Technology (CONACYT) for its financial support to with the "Cátedras–CONACYT 2015" program (No. 729) and by funding the scholarship program for Flores-García postgraduate studies (No. 710702). Author is indebted to Leslie Guzmán-Dávila for her assistance in the improvement of the figures showed in the present chapter.

CONFLICT OF INTEREST

The authors who have taken part in this assay declared that they do not have anything to disclose regarding funding or conflict of interest with respect to this manuscript.

KEYWORDS

- **adjuvant treatments**
- **health**
- **nutraceuticals**

REFERENCES

1. DeFelice, S. L., (1995). The nutraceutical revolution: Its impact on food industry R&D. *Trends Food Sci. Technol.*, *6*(2), 59–61.
2. Brower, V., (1998). Nutraceuticals: Poised for a healthy slice of the healthcare market? *Nat. Biotechnol.*, *16*(8), 728–31.
3. Rajasekaran, A., Sivagnanam, G., & Xavier, R., (2008). Nutraceuticals as therapeutic agents: A review. *Res. J. Pharm. Technol.*, *1*(4), 328–340.
4. Das, L., Bhaumik, E., Raychaudhuri, U., & Chakraborty, R., (2012). Role of nutraceuticals in human health. *J. Food Sci. Technol.*, *49*(2), 173–183.
5. Zeisel, S. H., (1999). Regulation of "nutraceuticals." *Science*, *285*(5435), 1853–1855.
6. Nasri, H., Baradaran, A., Shirzad, H., & Rafieian–Kopaei, M., (2014). New concepts in nutraceuticals as alternative for pharmaceuticals. *Int. J. Prev. Med.*, *5*(12), 1487–1499.
7. Shahidi, F., (2009). Nutraceuticals and functional foods: Whole versus processed foods. *Trends Food Sci. Technol.*, *20*(9), 376–387.
8. Nair, H. B., Sung, B., Yadav, V. R., Kannappan, R., Chaturvedi, M. M., & Aggarwal, B. B., (2010). Delivery of antiinflammatory nutraceuticals by nanoparticles for the prevention and treatment of cancer. *Biochem. Pharmacol.*, *80*(12), 1833–1843.
9. Chauhan, B., Kumar, G., Kalam, N., & Ansari, S. H., (2013). Current concepts and prospects of herbal nutraceutical: A review. *J. Adv. Pharm. Technol. Res.*, *4*(1), 4–8.
10. Scicchitano, P., Cameli, M., Maiello, M., Modesti, P. A., Muiesan, M. L., Novo, S. et al., (2014). Nutraceuticals and dyslipidaemia: Beyond the common therapeutics. *J. Funct. Foods*, *6*, 11–32.

11. Singh, J., & Sinha, S., (2012). Classification, regulatory acts and applications of nutraceuticals for health. *Int. J. Pharma Bio Sci.*, *2*, 177–187.
12. Lokesh, K. N., & Channarayappa V. M., (2015). Exemplified screening standardization of potent antioxidant nutraceuticals by principles of design of experiments. *J. Funct. Foods*, *17*, 260–270.
13. Pandey, M., Verma, R. K., & Saraf, S. A., (2010). Nutraceuticals: New era of medicine and health. *Asian J. Pharm. Clin. Res.*, *3*(1), 11–15.
14. Azmir, J., Zaidul, I. S. M., Rahman, M. M., Sharif, K. M., Mohamed, A., Sahena, F. et al., (2013). Techniques for extraction of bioactive compounds from plant materials: A review. *J. Food Eng.*, *117*(4), 426–436.
15. Gil-Chávez, G. J., Villa, J. A., Ayala-Zavala, J. F., Heredia, J. B., Sepulveda, D., Elhadi, M. Y. et al., (2013). Technologies for extraction and production of bioactive compounds to be used as nutraceuticals and food ingredients: An overview. *Compr. Rev. Food Sci. Food Saf.*, *12*(1), 5–23.
16. Wilken, L. R., & Nikolov, Z. L., (2012). Recovery and purification of plant-made recombinant proteins. *Biotechnol. Adv.*, *30*(2), 419–33.
17. Lai, S. M., & Gu, J. Y., (2014). Two-step chromatographic procedure for the preparative separation and purification of epigallocatechin gallate from green tea extracts. *Food Bioprod. Process.*, *92*(3), 314–320.
18. Mahabir, S., (2014). Methodological challenges conducting epidemiological research on nutraceuticals in health and disease. *Pharma. Nutrition*, *2*(3), 120–125.
19. Bian, W. P., & Pei, D. S., (2016). Zebrafish model for safety and toxicity testing of nutraceuticals. In: *Nutraceuticals: Efficacy, Safety and Toxicity*, Elsevier Science & Technology San Diego, USA.
20. Krishna, G., & Gopalakrishnan, G., (2016). Alternative *in vitro* models for safety and toxicity evaluation of nutraceuticals. In: *Nutraceuticals: Efficacy, Safety and Toxicity*, p. 355.
21. Gonzalez-Suarez, I., Martin, F., Hoeng, J., & Peitsch, M. C., (2016). Mechanistic network models in safety and toxicity evaluation of nutraceuticals. In: *Nutraceuticals: Efficacy, Safety and Toxicity*, p. 287.
22. Jalbert, I., (2013). Diet, nutraceuticals and the tear film. *Exp. Eye Res.*, *117*, 138–146.
23. Anbazahan, S., Harikrishnan, R., & Jawahar, S., (2014). Nutraceutical studies in *Morinda citrifolia* linn fruit. *IRJES, Int. Ref. J. Eng. Sci.*, *3*(6), 60–63.
24. Je, J. Y., Park, S. Y., Hwang, J. Y., & Ahn, C. B., (2015). Amino acid composition and *in vitro* antioxidant and cytoprotective activity of abalone viscera hydrolysate. *J. Funct. Foods*, *16*, 94–103.
25. Vijay, S., Lalit, S., Navneet, V., & Garima, K., (2016). "The nutraceutical amino acids"- nature's fortification for robust health. *Br. J. Pharm. Res.*, *11*(3), 1–20.
26. Valiño, V., San Román, M. F., Ibáñez, R., Benito, J. M., Escudero, I., & Ortiz, I., (2014). Accurate determination of key surface properties that determine the efficient separation of bovine milk BSA and LF proteins. *Sep. Purif. Technol.*, *135*, 145–157.
27. Kimpel, F., & Schmitt, J. J., (2015). Review: Milk proteins as nanocarrier systems for hydrophobic nutraceuticals. *J. Food Sci.*, *80*(11), 2361–2366.
28. Diarrassouba, F., Remondetto, G., Garrait, G., Alvarez, P., Beyssac, E., & Subirade, M., (2015). Self-assembly of β-lactoglobulin and egg white lysozyme as a potential carrier for nutraceuticals. *Food Chem.*, *173*, 203–209.

29. Nieves, J. W., (2013). Skeletal effects of nutrients and nutraceuticals, beyond calcium and vitamin D. *Osteoporosis Int.*, *24*(3), 771–786.
30. Houston, M. C., (2013). The role of nutrition, nutraceuticals, vitamins, antioxidants, and minerals in the prevention and treatment of hypertension. *Altern. Ther. Health Med.*, *19*(1), 32–49.
31. Webster, G. A., & Lehrke, P., (2013). Development of a combined bovine colostrum and immune-stimulatory carbohydrate nutraceutical for enhancement of endogenous stem cell activity. *Open Nutraceuticals J.*, *6*(1), 35–44.
32. Pallela, R., (2014). Chapter 9 - Nutraceutical and pharmacological implications of marine carbohydrates. In: Se-Kwon, K., (ed.), *Advances in Food and Nutrition Research* (Vol. 73, pp. 183–195). Academic Press.
33. Bahadoran, Z., Mirmiran, P., & Azizi, F., (2013). Dietary polyphenols as potential nutraceuticals in management of diabetes. A review. *J. Diabetes Metab. Disord.*, *12*(1), 43.
34. López-Gutiérrez, N., Romero-González, R., Martínez Vidal, J. L., & Frenich, A. G., (2016). Determination of polyphenols in grape-based nutraceutical products using high resolution mass spectrometry. *LWT--Food Sci. Technol.*, *71*, 249–259.
35. Tomé-Carneiro, J., & Visioli, F., (2016). Polyphenol-based nutraceuticals for the prevention and treatment of cardiovascular disease: Review of human evidence. *Phytomedicine*, 1–30.
36. Alanazi, A. S., (2013). The role of nutraceuticals in the management of autism. *Saudi Pharm. J.*, *21*(3), 233–43.
37. Hamilton, M. L., Haslam, R. P., Napier, J. A., & Sayanova, O., (2014). Metabolic engineering of *Phaeodactylum tricornutum* for the enhanced accumulation of omega–3 long chain polyunsaturated fatty acids. *Metab. Eng.*, *22*, 3–9.
38. Ganguly, S., (2013). Nutraceutical and pharmaceutical implication of prebiotics in livestock and poultry feed. *Bull. Pharm. Res.*, *3*(2), 71–77.
39. Samanta, A. K., Jayapal, N., Jayaram, C., Roy, S., Kolte, A. P., Senani, S. et al., (2015). Xylooligosaccharides as prebiotics from agricultural by-products: Production and applications. *Bioact. Carbohydr. Diet. Fibre*, *5*(1), 62–71.
40. Wichienchot, S., Youravong, W., Prueksasri, S., & Ngampanya, B., (2015). Recent researches on prebiotics for gut health in Thailand. *Funct. Foods Health Dis.*, *5*(11), 381–394.
41. Behnsen, J., Deriu, E., Sassone-Corsi, M., & Raffatellu, M., (2013). Probiotics: Properties, examples, and specific applications. *Cold Spring Harbor Perspect. Med.*, *3*(3), a010074.
42. Begum, S. B., Roobia, R. R., Karthikeyan, M., & Murugappan, R. M., (2015). Validation of nutraceutical properties of honey and probiotic potential of its innate microflora. *LWT--Food Sci. Technol.*, *60*(2, Part 1), 743–750.
43. Trung, T. S., & Stevens, W. F., (2013). Extraction of nutraceuticals from shrimp by-products. In: *Marine Nutraceuticals: Prospects and Perspectives* (p. 115).
44. Tanaka, T., Shnimizu, M., & Moriwaki, H., (2012). Cancer chemoprevention by carotenoids. *Molecules (Bas, Switz)*, *17*(3), 3202–3242.
45. Berrow, E. J., Bartlett, H. E., Eperjesi, F., & Gibson, J. M., (2013). The effects of a lutein-based supplement on objective and subjective measures of retinal and visual

function in eyes with age-related maculopathy -- a randomised controlled trial. *Br. J. Nutr.*, *109*(11), 2008–2014.

46. Gonzalez-Sarrias, A., Larrosa, M., Garcia-Conesa, M. T., Tomas-Barberan, F. A., & Espin, J. C., (2013). Nutraceuticals for older people: Facts, fictions and gaps in knowledge. *Maturitas, 75*(4), 313–34.
47. Chan, A. T., & Giovannucci, E. L., (2010). Primary prevention of colorectal cancer. *Gastroenterol.*, *138*(6), 2029–2043.
48. World Health Organization. Cancer: description note 2017. http://www.who.int/mediacentre/factsheets/fs297/es/.
49. DeSantis, C. E., Lin, C. C., Mariotto, A. B., Siegel, R. L., Stein, K. D., Kramer, J. L. et al., (2014). Cancer treatment and survivorship statistics. *CA, Cancer J. Clin.*, *64*(4), 252–271.
50. Lo Russo, G., Proto, C., & Garassino, M. C., (2016). Afatinib in the treatment of squamous non-small cell lung cancer: A new frontier or an old mistake? *Transl. Lung Cancer Res.*, *5*(1), 110–114.
51. Rudin, C. M., Liu, W., Desai, A., Karrison, T., Jiang, X., Janisch, L. et al., (2008). Pharmacogenomic and pharmacokinetic determinants of erlotinib toxicity. *J. Clin. Oncol.*, *26*(7), 1119–1127.
52. Cranganu, A., & Camporeale, J., (2009). Nutrition aspects of lung cancer. *Nutr. Clin. Pract.*, *24*(6), 688–700.
53. Ku, M. J., Kim, J. H., Lee, J., Cho, J. Y., Chun, T., & Lee, S. Y., (2015). Maclurin suppresses migration and invasion of human non-small-cell lung cancer cells via anti-oxidative activity and inhibition of the Src/FAK–ERK–β-catenin pathway. *Mol. Cell. Biochem.*, *402*(1), 243–252.
54. Wright, M. E., Mayne, S. T., Stolzenberg-Solomon, R. Z., Li, Z., Pietinen, P., Taylor, P. R. et al., (2004). Development of a comprehensive dietary antioxidant index and application to lung cancer risk in a cohort of male smokers. *Am. J. Epidemiol.*, *160*(1), 68–76.
55. Frese, S., Frese-Schaper, M., Andres, A. C., Miescher, D., Zumkehr, B., & Schmid, R. A., (2006). Cardiac glycosides initiate Apo2L/TRAIL-induced apoptosis in non–small cell lung cancer cells by up-regulation of death receptors 4 and 5. *Cancer Res.*, *66*(11), 5867–5874.
56. Katiyar, S. K., Meeran, S. M., Katiyar, N., & Akhtar, S., (2009). p53 Cooperates berberine-induced growth inhibition and apoptosis of non-small cell human lung cancer cells in vitro and tumor xenograft growth *in vivo. Mol. Carcinog.*, *48*(1), 24–37.
57. Abou-Alfa, G. K., & DeMatteo, R., (2011). *100 Questions & Answers About Liver Cancer*. Jones & Bartlett Learning.
58. Oishi, N., Yamashita, T., & Kaneko, S., (2014). Molecular biology of liver cancer stem cells. *Liver Cancer*, *3*(2), 71–84.
59. Mittal, S., & El-Serag, H. B., (2013). Epidemiology of HCC: Consider the population. *J. Clin. Gastroenterol.*, *47*(0), 2–6.
60. Bruix, J., Han, K. H., Gores, G., Llovet, J. M., & Mazzaferro, V., (2015). Liver cancer: Approaching a personalized care. *J. Hepatol.*, *62*(1, Supplement), 144–156.
61. Greten, T. F., Wang, X. W., & Korangy, F., (2015). Current concepts of immune based treatments for patients with HCC: From basic science to novel treatment approaches. *Gut*, *64*(5), 842–848.

62. Sethi, G., Chatterjee, S., Rajendran, P., Li, F., Shanmugam, M. K., Wong, K. F., Kumar, A. P. et al., (2014). Inhibition of STAT3 dimerization and acetylation by garcinol suppresses the growth of human hepatocellular carcinoma *in vitro* and *in vivo*. *Mol. Cancer.*, *13*, 66–66.
63. Michailidou, M., Melas, I. N., Messinis, D. E., Klamt, S., Alexopoulos, L. G., Kolisis, F. N. et al., (2015). Network-based analysis of nutraceuticals in human hepatocellular carcinomas reveals mechanisms of chemopreventive action. *CPT: Pharmacometrics Syst. Pharmacol.*, *4*(6), 350–361.
64. World Health Orgnization. Colorectal cancer, (2017). http://www.paho.org/hq/index.php?option=com_content&view=article&id=11761%3Acolorectal-cancer-&catid=1872%3Acancer&Itemid=41765&lang=es.
65. Jansen, L., Koch, L., Brenner, H., & Arndt, V., (2010). Quality of life among long-term (>/=5 years) colorectal cancer survivors—systematic review. *Eur. J. Cancer*, *46*(16), 2879–2888.
66. Kuppusamy, P., Yusoff, M. M., Maniam, G. P., Ichwan, S. J. A., Soundharrajan, I., & Govindan, N., (2014). Nutraceuticals as potential therapeutic agents for colon cancer: A review. *Acta Pharm. Sin. B*, *4*(3), 173–181.
67. Chen, J., (2012). Prevention of obesity-associated colon cancer by (-)-epigallocatechin-3 gallate and curcumin. *Transl. Gastrointest. Cancer*, *1*(3), 243–249.
68. Kaur, M., Tyagi, A., Singh, R. P., Sclafani, R. A., Agarwal, R., & Agarwal, C., (2011). Grape seed extract upregulates p21 (Cip1) through redox-mediated activation of ERK1/2 and posttranscriptional regulation leading to cell cycle arrest in colon carcinoma HT29 cells. *Mol. Carcinog.*, *50*(7), 553–562
69. Khanal, P., Kang, B. S., Yun, H. J., Cho, H. G., Makarieva, T. N., & Choi, H. S., (2011). Aglycon of rhizochalin from the *Rhizochalina incrustata* induces apoptosis via activation of AMP-activated protein kinase in HT–29 colon cancer cells. *Biol. Pharm. Bull.*, *34*(10), 1553–1558.
70. Carrera-González, M. P., Ramírez-Expósito, M. J., Mayas, M. D., & Martínez-Martos, J. M., (2013). Protective role of oleuropein and its metabolite hydroxytyrosol on cancer. *Trends Food Sci. Technol.*, *31*(2), 92–99.
71. Karimi, P., Islami, F., Anandasabapathy, S., Freedman, N. D., & Kamangar, F., (2014). Gastric cancer: Descriptive epidemiology, risk factors, screening, and prevention. *Cancer Epidemiol. Biomarkers Prev.*, *23*(5), 700–713.
72. Qiao, Y. L., Dawsey, S. M., Kamangar, F., Fan, J. H., Abnet, C. C., Sun, X. D. et al., (2009). Total and cancer mortality after supplementation with vitamins and minerals: follow-up of the linxian general population nutrition intervention trial. *J. Natl. Cancer Inst.*, *101*(7), 507–518.
73. Serafini, M., Jakszyn, P., Lujan-Barroso, L., Agudo, A., Bas Bueno-de-Mesquita, H., Van Duijnhoven, F. J. et al., (2012). Dietary total antioxidant capacity and gastric cancer risk in the European prospective investigation into cancer and nutrition study. *Int. J. Cancer*, *131*(4), 544–554.
74. Ullah, M. F., Bhat, S. H., Husain, E., Abu-Duhier, F., Hadi, S. M., Sarkar, F. H., & Ahmad, A., (2016). Pharmacological intervention through dietary nutraceuticals in gastrointestinal neoplasia. *Crit. Rev. Food Sci. Nutr.*, *56*(9), 1501–18.

75. Torre, L. A., Bray, F., Siegel, R. L., Ferlay, J., Lortet-Tieulent, J., & Jemal, A., (2015). *CA Cancer J. Clin. CA: A Cancer Journal for Clinicians*, *65*(2), 87–108.
76. Gartner, R., Jensen, M. B., Nielsen, J., Ewertz, M., Kroman, N., & Kehlet, H., (2009). Prevalence of and factors associated with persistent pain following breast cancer surgery. *JAMA, J. Am. Med. Assoc.*, *302*(18), 1985–1992.
77. Vilholm, O. J., Cold, S., Rasmussen, L., & Sindrup, S. H., (2008). The postmastectomy pain syndrome: an epidemiological study on the prevalence of chronic pain after surgery for breast cancer. *Br. J. Cancer*, *99*(4), 604–610.
78. Conte, P., & Frassoldati, A., (2007). Aromatase inhibitors in the adjuvant treatment of postmenopausal women with early breast cancer: Putting safety issues into perspective. *Breast J.*, *13*(1), 28–35.
79. Pinto, A. C., & De Azambuja, E., (2011). Improving quality of life after breast cancer: Dealing with symptoms. *Maturitas*, *70*(4), 343–348.
80. Motaghed, M., Al-Hassan, F. M., & Hamid, S. S., (2013). Cellular responses with thymoquinone treatment in human breast cancer cell line MCF–7. *Pharmacogn. Res.*, *5*(3), 200–206.
81. Mock, C. D., Jordan, B. C., & Selvam, C., (2015). Recent advances of curcumin and its analogues in breast cancer prevention and treatment. *RSC Adv.*, *5*(92), 75575–75588.
82. He, F.J., & Chen, J.Q., (2013). Consumption of soybean, soy foods, soy isoflavones and breast cancer incidence: Differences between Chinese women and women in western countries and possible mechanisms. *Food Sci. Hum. Wellness*, *2*, 146–161.
83. Alzheimer's Association, (2015). Alzheimer's disease facts and figures. *Alzheimer's & Dementia: the Journal of the Alzheimer's Association*, *11*(3), 332–384.
84. Reitz, C., & Mayeux, R., (2014). Alzheimer disease. Epidemiology, diagnostic criteria, risk factors and biomarkers. *Biochem. Pharmacol.*, *88*(4), 640–651.
85. Pringsheim, T., Jette, N., Frolkis, A., & Steeves, T. D., (2014). The prevalence of Parkinson's disease: a systematic review and meta-analysis. *Mov. Disord.*, *29*(13), 1583–1590.
86. Connolly, B. S., & Lang, A. E., (2014). Pharmacological treatment of parkinson disease: A review. *JAMA, J. Am. Med. Assoc.*, *311*(16), 1670–1683.
87. Akhondzadeh, S., Sabet, M. S., Harirchian, M. H., Togha, M., Cheraghmakani, H., Razeghi, S. et al., (2010). Saffron in the treatment of patients with mild to moderate Alzheimer's disease: A 16-week, randomized and placebo-controlled trial. *J. Clin. Pharm. Ther.*, *35*(5), 581–588.
88. Polidori, M. C., Pratico, D., Mangialasche, F., Mariani, E., Aust, O., Anlasik, T. et al., (2009). High fruit and vegetable intake is positively correlated with antioxidant status and cognitive performance in healthy subjects. *J. Alzheimer's Dis.*, *17*(4), 921–927.
89. Kelsey, N. A., Wilkins, H. M., & Linseman, D. A., (2010). Nutraceutical antioxidants as novel neuroprotective agents. *Molecules (Bas, Switz)*, *15*(11), 7792–814.
90. Kim, D., Nguyen, M. D., Dobbin, M. M., Fischer, A., Sananbenesi, F., Rodgers, J. T. et al., (2007). SIRT1 deacetylase protects against neurodegeneration in models for Alzheimer's disease and amyotrophic lateral sclerosis. *EMBO J.*, *26*(13), 3169–3179.
91. Drukarch, B., Flier, J., Jongenelen, C. A., Andringa, G., & Schoffelmeer, A. N., (2006). The antioxidant anethole dithiolethione inhibits monoamine oxidase-B but

not monoamine oxidase A activity in extracts of cultured astrocytes. *J. Neural Transm. (Vienna)*, *113*(5), 593–598.
92. Radad, K., Moldzio, R., Taha, M., & Rausch, W. D., (2009). Thymoquinone protects dopaminergic neurons against MPP+ and rotenone. *Phytother. Res.*, *23*(5), 696–700.
93. Kim, S. J., Kim, J. S., Cho, H. S., Lee, H. J., Kim, S. Y., Kim, S. et al., (2006). Carnosol, a component of rosemary (*Rosmarinus officinalis* L.) protects nigral dopaminergic neuronal cells. *Neuroreport*, *17*(16), 1729–1733.
94. Filomeni, G., Graziani, I., De Zio, D., Dini, L., Centonze, D., Rotilio, G. et al., (2012). Neuroprotection of kaempferol by autophagy in models of rotenone-mediated acute toxicity: Possible implications for Parkinson's disease. *Neurobiol. Aging*, *33*(4), 767–785.
95. Copeland, R. A., (2004). *Enzymes: A Practical Introduction to Structure, Mechanism, and Data Analysis* (2 edn.), VCH Publishers New York.
96. FitzGerald, R. J., & Meisel, H., (2000). Milk protein-derived peptide inhibitors of angiotensin-i-converting enzyme. *Br. J. Nutr.*, *84* (1), 33–37.
97. Moncada-Pazos, A., Obaya, A. J., Viloria, C. G., Lopez-Otin, C., & Cal, S., (2011). The nutraceutical flavonoid luteolin inhibits ADAMTS–4 and ADAMTS–5 aggrecanase activities. *J. Mol. Med. (Berl)*, *89*(6), 611–619.
98. Faisal, N., & Varma, K. S., (2006). Nutraceuticals and its impact on health care. *B. Pharm. Projects and Review Articles*, *1*, 299–331.
99. Menon, V. V., & Lele, S. S., (2015). Nutraceuticals and bioactive compounds from seafood processing waste. In: Kim, S. K., (ed.), *Springer Handbook of Marine Biotechnology* (pp. 1405–1425). Springer, Berlin Heidelberg: Berlin, Heidelberg.
100. Díaz-López, M., & García-Carreño, F. L., (2000). Applications of fish and shellfish enzymes in food and feed products. In: Haard N. F. S. B. K., (eds.), *Seafood Enzymes* (pp. 571–618). New York, NY: Marcel Dekker.
101. Shih, P. H., Hwang, S. L., Yeh, C. T., & Yen, G. C., (2012). Synergistic effect of cyanidin and PPAR agonist against nonalcoholic steatohepatitis-mediated oxidative stress-induced cytotoxicity through MAPK and Nrf2 transduction pathways. *J. Agric. Food Chem.*, *60*(11), 2924–2933.
102. Nijveldt, R. J., Van Nood, E., Van Hoorn, D. E., Boelens, P. G., Van Norren, K., & Van Leeuwen, P. A., (2001). Flavonoids: A review of probable mechanisms of action and potential applications. *Am. J. Clin. Nutr.*, *74*(4), 418–425.
103. Campolongo, G., Riccioni, C. V., Raparelli, V., Spoletini, I., Marazzi, G., Vitale, C. et al., (2016). The combination of nutraceutical and simvastatin enhances the effect of simvastatin alone in normalising lipid profile without side effects in patients with ischemic heart disease. *IJC Metab. Endoc.*, *11*, 3–6.
104. Pan, M. H., Lai, C. S., Wu, J. C., & Ho, C. T., (2013). Effect of flavonoids from fruits and vegetables in the prevention and treatment of cancer. In: Springer, *Cancer Chemoprevention and Treatment by Diet Therapy* (pp. 23–54).
105. Davinelli, S., Calabrese, V., Zella, D., & Scapagnini, G., (2014). Epigenetic nutraceutical diets in alzheimer's disease. *J. Nutr. Health Aging*, *18*(9), 800–805.
106. Rasmussen, H. E., Piazza, B. R., Forsyth, C. B., & Keshavarzian, A., (2014). Nutrition and gastrointestinal health as modulators of Parkinson's disease. In: *Pharma-Nutrition*, Springer, pp. 213–242.

107. Sahebkar, A., Serban, M. C., Gluba-Brzózka, A., Mikhailidis, D. P., Cicero, A. F., Rysz, J. et al., (2016). Lipid-modifying effects of nutraceuticals: An evidence-based approach. *Nutrition. 32*, 1179–1192.
108. Roberfroid, M., (2007). Prebiotics: The concept revisited. *J. Nutr.*, *137*(3 Suppl. 2), 830–837.
109. Katia, P., & Chiara, T., (2015). The role of nutraceuticals. In: *Connecting Indian Wisdom and Western Science* (pp. 101–118), CRC Press.
110. Laparra, J. M., & Sanz, Y., (2010). Interactions of gut microbiota with functional food components and nutraceuticals. *Pharmacol. Res.*, *61*(3), 219–225.
111. De Vrese, M., & Schrezenmeir, J., (2008). Probiotics, prebiotics, and synbiotics. *Adv. Biochem. Eng./Biotechnol.*, *111*, 1–66.
112. Anadón, A., Martínez-Larrañaga, M. R., Ares, I., & Martínez, M. A., (2016). Chapter 54 - Prebiotics: Safety and toxicity considerations. In: *Nutraceuticals* (pp. 757–775). Academic Press. Boston.
113. Akhtar, S., Ismail, T., Fraternale, D., & Sestili, P., (2015). Pomegranate peel and peel extracts: Chemistry and food features. *Food Chem.*, *174*, 417–425.
114. Butel, M. J., (2014). Probiotics, gut microbiota and health. *Med. Mal. Infect*, *44*(1), 1–8.
115. Gupta, C., & Prakash, D., (2015). Nutraceuticals for geriatrics. *J. Tradit. Complement. Med.*, *5*(1), 5–14.
116. Jensen, S. J. K., (2003). Oxidative stress and free radicals. *J. Mol. Struct.: THEOCHEM.*, *666*, 387–392.
117. Valko, M., Rhodes, C. J., Moncol, J., Izakovic, M., & Mazur, M., (2006). Free radicals, metals and antioxidants in oxidative stress-induced cancer. *Chem. Biol. Interact.*, *160*(1), 1–40.
118. Valko, M., Leibfritz, D., Moncol, J., Cronin, M. T., Mazur, M., & Telser, J., (2007). Free radicals and antioxidants in normal physiological functions and human disease. *Int. J. Biochem. Cell Biol.*, *39*(1), 44–84.
119. Fraga, C. G., Galleano, M., Verstraeten, S. V., & Oteiza, P. I., (2010). Basic biochemical mechanisms behind the health benefits of polyphenols. *Mol. Aspects Med.*, *31*(6), 435–445.
120. Prasad, S., Gupta, S. C., & Tyagi, A. K., (2017). Reactive oxygen species (ROS) and cancer: Role of antioxidative nutraceuticals. *Cancer Lett.*, *387*, 95–105.
121. Brunetti, C., Di Ferdinando, M., Fini, A., Pollastri, S., & Tattini, M., (2013). Flavonoids as antioxidants and developmental regulators: Relative significance in plants and humans. *Int. J. Mol. Sci.*, *14*(2), 3540–3555.

CHAPTER 7

TAGETES LUCIDA: A RELATIONSHIP BETWEEN COMPOUNDS AND PROPERTIES FOR THE DEVELOPMENT OF FUNCTIONAL FOODS

PERLA YANETH VILLA SILVA, CRYSTEL ALEYVICK SIERRA RIVERA, LLUVIA ITZEL LÓPEZ LÓPEZ, ANNA ILINÁ, JUAN ALBERTO ASCACIO VALDÉS, and SONIA YESENIA SILVA BELMARES

Food Research Department, Faculty of Chemistry, Autonomous University of Coahuila, Blvd. Venustiano Carranza, Colony Republic, Zip code 25280, Saltillo, Coahuila, Mexico, E-mail: yesenia_silva@uadec.edu.mx

ABSTRACT

Tagetes lucida is endemic to Mexico and Central Region of America, it is known as Hierbanís and it is a natural source of metabolites that could be exploited for the development of functional foods. In Mexico and in the world, foods do not only provide nutrients for humans, but they also prevent diseases and bring health benefits. This manuscript offers information that allows to identify new perspectives for the development of functional foods using as natural source *T. lucida*. In this review, health benefits that could be provided from the plants of the steraceae family including *T. lucida* are identified, through the elaboration of food products such as juices or solid foods. These foods could have different actions according to the included active isolated principles incorporated.

7.1 INTRODUCTION

At present, the consumption of functional foods is of interest, and as a result, many foods include plant extracts with beneficial properties for human health [1–3]. A functional food maintains the organoleptic characteristics and improves health conditions due to its nutrients, diminishing the risk of some diseases [4].

Since ancient times, it has been known that plants have therapeutic properties and in recent years it has been found that they contain a wide range of compounds [5]. These scientific advances have allowed the development of new food products whose content includes bioactive compounds [6, 7].

The tagetes genus includes 56 species [8–9], belongs to the asteraceae family and includes species with biological properties. *T. erecta* [10], *T. patula* [11], *T. minuta* [12], *T. rupestris, T. terniflora* [13] and *Tagetes lucida* [8] are some examples. Other species of the astaeraceae family show anti-inflammatory and anticancer activities as well [14]. The *Tanacetum vulgare* L. (Asteraceae) essential oil shows effects against *Escherichia coli* and *Staphylococcus aureus* [15].

T. erecta and *T. patula* L. show many activities such as antimicrobial, antioxidant, antidiabetic, and antilipemic [16, 17]. Additionally, the *T. patula* L. flowers can be used to produce a lutein ester, which is a potential food supplement [18]. *T. minuta* essential oil has an anti-inflammatory effect and the ability to eliminate radicals proving to have antioxidant activity [12]. Additionally, this plant shows a cytotoxic effect against MCF–7 breast tumor cells and antifungal properties [19]. *T. rupestris* shows repellent properties against *Ceratitis capitata,* so it could be used to avoid deterioration of fruits and reduce economic losses [13].

7.2 BOTANICAL APPROACH AND DISTRIBUTION

T. lucida is a Mexican plant known as Hierbanis, pericón, Santa María, Mexican tarragon and sweet dragon. *T. lucida,* is distributed in Mexico and Central America [22, 23]. The plant grows in places with a moderate climate, it has a pleasant aroma, a total height of 80 centimeters and branched stem. Additionally, some of the morphological features of this plant are simple leaves with dimensions of 2 to 10 cm of length and 0.5 to

2 cm of width. The flowers are yellow-gold or yellow-orange with a diameter of 1–1.5 cm [24–27]. This plant has antibacterial, antifungal, antimalarial and antidepressant activities [8, 9, 28].

According to the information described previously, *T. lucida* has beneficial health properties, so it could be a source of functional foods. Additionally, tagetes plants contain compounds that provide these biological activities, being promising for the development of new food products [6, 7].

7.3 COMPOUNDS OF IMPORTANCE FOR FOOD DEVELOPMENT

Phytochemicals compounds are produced from plants; therefore, they are contained in many foods. They are classified according to their chemical structure and biological activity [29–32]. Some authors consider phytochemical compounds as secondary metabolites. Others state that these compounds have both beneficial and toxic effects, which is relevant for scientific investigation in the evaluation of the biological activity and identification of the active compounds [33–36].

The phytochemical compounds extraction can be carried out with solvents of different polar nature. However, the most used in the food industry is ethanol, since it has low toxicity and is detected at traceslevel [37, 38]. Recently, Asteraceae compounds that have biological activity have been isolated and it has been found that they have potential for functional foods development. Some examples are 3β-carboxylic-4 (23)-ene, adian-5-en-3α-ol, fernenol, and fern-7-3β-ol, which were isolated from *Ainsliaea yunnanensis* Franch. These compounds showed a cytotoxic effect against a cell line and human acute monocytic leukemia (THP–1) [39]. *Smallanthus sonchifolius* contains fructooligosaccharides that incorporated to dietary supplements could prevent chronic diseases. They are known for favoring the growth of beneficial to health bacteria, reducing the pathogenic bacterial population [40]. Another example is the steviol glycoside sweetener, flavonoid chlorogenic acids and glycosides contained in *Stevia rebaudiana* leaves. This plant has a great economic importance and it is used in the processed foods industry as a sweetener additive. *Gnaphalium affine* contains quercetin as its main component, this compound inhibits the oxidation of peanut oil and lard, and it is used to prepare functional foods and nutraceuticals [41, 75].

As described above, *Tagetes lucida* has compounds with promising biological activities, an example of this is dimethylfraxetine, which has anxiolytics and sedatives properties [42]. This plant also contains flavonoids, steroids, alkaloids, and coumarins. These compounds have health benefits; therefore, the plant extracts could be used for the development of functional foods [43, 44]. Some phytochemical compounds, properties, and functions from Asteraceae family are shown in Table 7.1.

TABLE 7.1 Phytochemical Compounds, Biological Properties, and Asteraceae Function

Active compound	Property	Function/benefit	Reference
Terpene	Antioxidant	Cancer prevention	[45]
Sterol	Anti-inflammatory	Reduction of inflammation	[76, 77]
Phenol	Antioxidant, antimicrobial	Prevention of cardiovascular disease	[78, 79]
Flavonoid	Antitumor	Enzyme inhibition	[80, 81]
Coumarine	Biological and pharmacological	Prevents blood clotting	[82]
Anthocyanin	Antioxidant	Prevents cell degeneration	[83, 84]
Alkaloids	Hypoglycemic Antispasmodic and antidiarrheal activities	Decrease intestinal motility	[85]
Saponin	Antiviral and cytotoxic activities	Prevents the stomach and intestine cancers	[86]

7.3.1 FUNCTIONALITY OF PHYTOCHEMICAL COMPOUNDS

The active ingredients integrated into the functional food are present in a large number of plants, and incorporated to foods of natural origin or processed foods could provide a health-promoting effect [45]. There is a close relationship between the antioxidant and antibacterial effects shown by some species of Tagetes. The antioxidant effect has been detected by the 2,2-diphenyl-1-picrylhydrazyl (DPPH), 2,2-azino-bis–3-ethylbenzothiazoline–6-sulfonic acid (ABTS) and the lipid peroxyl (LP) test. This result has been observed on the plants *T. minuta and T. lucida* [46, 47].

T. lucida has a great nutritional and medicinal value, since it has antimicrobial activity against pathogenic enterobacteria such as enteropathogenic

Escherichia coli, Salmonella enteritidis, Salmonella typhi, Shigella dysenteriae, and *Shigella flexneri* [48].

In *Tagetes lucida,* a 7-methoxy-coumarin and 6,7-di-methoxy-coumarin shows toxicity in *A. salina.* This crustacean is used to evaluate preliminary food functionality [49, 50]. Other compounds detected in *T. lucida* are quercetagenin, 3,4-dimethyl-ether–7-ObD-glucopyranoside, and two new phenolic acids: 3–2-ObD-glucopyranosyl-4-methoxy-phenyl) -propanoic acid and methyl ester [47]. On the other hand, a research has detected an antidepressant effect on *T. lucida* through quercentin [51]. There are also anxiolytic and sedative effects reported from dimethylfraxtetin isolated from this plant [52]. *T. lucida* contains apigenin and chlorogenic, these phenolic acids have health benefits since they are easily absorbed through the intestinal tract and functions as an antioxidant [53].

As it has been mentioned, the genus targets contain compounds with various beneficial activities, however, many of them have not been reported. *Tagetes lucida* [54] and *T. patula* contain quercetin, flavonoids, carotenoids, and phenolic compounds [16]. The patuletin showed anti-inflammatory and anti-arthritic properties in a model of arthritic rodents, so it is considered as a possible immunosuppressive and antiarthritic treatment [11]. *T. erecta* contains lutein [55] and carotenoids therefore it could be used as a nutritional supplement on humans and animals, since carotenoids are part of some birds food [56].

T. lucida contains compounds such as alkaloids, phenolic compounds, flavonoids, coumarins, anthocyanins, terpenes, and steroids; some of them are found on other Asteraceae species [22, 23, 57]. In addition to this, a component of the essential oils of this plant has been identified as methyl-chavicol [58], along with this compound bithienyls has been detected as a minor component [59]. The essential oil of *T. lucida* contains Linalool and methyl-chavicol. *T. erecta, T. glandulifera, T. signata, T. tenuifolia* and *T. temuifolia* contain p-ocimene, limonene, a-terpinolene, dihydroxyacetone, artemisia ketone, thujone, tagetone, camphor, umbellulone, 2-ocimenone, E-octaimenone, piperitone, metyleugenol, P-caryophyllene, and piperitenone [46, 54].

Additionally, in *Tagetes caracasana* were identified trans-odenona and cis-tagetone, in *T. erecta* piperitone and terpinolene, in *Tagetes filifolia* trans-anetol and estragol, in *T. subulata* terpinolene, piperitenone, and limonene, while in *T. patula* were identified terpinolene, piperitenone, b-caryophyllene, terpinolene, and cis-b-ocimene [60].

On the other hand, compounds of promising biological activities were detected in asteraceae plants. One of them is the arctigenin that has anti-inflammatory and anti-carcinogenic activities [14]. The flavonoids, steroids, alkaloids, and coumarins form Asteraceae species provide health benefits [43, 44]. In *Verbesina encelioides* (asteraceae), Pseudotaraxaxolol–3β-acetate, 16β-hydroxy-pseudotaraxasterol–3β-palmitate, sitosterol β-glucoside and β-sitosterol galactoside have been identified with antiprotozoal effects [61].

Monogalactosyldiacylglycerol from *Cirsium brevicaule* A. GRAY decreases hepatic lipid accumulation and the expression of the fatty acid synthase (FASN) gene in mice, having potential as a food ingredient [62]. The biggest genus of the asteraceae family is Artemisia; in the essential oils of some species, there are 1, 8-cineol, beta-pinene, tuyona, artemisia ketone, camphor, caryophyllene, camphene, and germacrene D. This genus has antimicrobial, insecticides, and antioxidants activities [63].

The essential oil of *Tanacetum vulgare* L. contains humulene with anti-inflammatory activity, pinene, and caryophyllene oxide have antioxidant activity, while camphor, as well as caryophyllene oxide, present an antibacterial effect. Also, humulene and caryophyllene oxide were found to be moderately cytotoxic against A–549, DLD–1, and WS1 cell lines [64]. *Calendula stellata* contains saponins with effect on *Staphylococcus auresus* and *Enterococcus faecalis* analogous to conventional antibiotics. Furthermore, it has cytotoxic activity against the fibrosarcoma cell lines (HT1080) and human lung cancer (A549) [65].

As it has been described, the family asteraceae has an immeasurable number of compounds and biological activities that are related to the members of this group of plants. These discoveries prove that *Tagetes lucida* could be used as an additive with antibacterial and antioxidant properties, to protect food from oxidative damage and food-borne pathogens.

7.3.2 DEVELOPMENT OF FUNCTIONAL FOOD

At present, many people are insecure when consuming some food, since many of them do not have enough scientific studies to validate a good nutritional quality. As a result, there is an urgent need to carry out programs that allow farmers and food systems to be sensitive to nutrition and health care [66]. In recent years, the growing use of nutraceuticals and derived products of plants in developed countries has been witnessed [67].

Currently, food supplements made from plant species are used in traditional medicine. In many cases, the collection, taxonomic identification of the species, techniques of extraction, standardization, product homogenization, pharmacological, and clinical tests, are managed at free will. This is because consumers of herbal remedies believe that all herbal products are safe and effective, so, they prefer to treat themselves with products that lack sanitary regulation [68]. Although many plants have nutritional and beneficial effects for health, they must be studied to ensure their effectiveness and safety.

Accordingly, some companies have developed technologies to produce functional foods with scientific validity. One example is the development of products from wheat bran, which is an abundant by-product of the milling industry, full of bioactive compounds and fat-soluble fibers. Some of its bioactive compounds are oryzanol, tocopherols, and carotenoids, all of these compounds have an antioxidant effect in the liver. Therefore, wheat bran has a high potential as an ingredient in the formulation of healthy foods. It is worth mentioning that some of the compounds found in this plant are found in Asteraceae species [69].

Other studies have researched the release of hydrogen cyanide from cyanogenic glycosides of forage plants (*Sorghum spp.*, *Trifolium spp.*, and *Lotus spp.*), since its consumption represents a serious problem for the animals [70].

The development of these researches has brought knowledge on the nutritional advantages and the innocuity on new alimentary products. For this reason, in recent times, the need to incorporate new chemical and nutritional compounds in food has increased. Hence, the isolation of phytochemicals compounds offers new opportunities for the development of food products [71, 72]. According to existing information, there is insufficient data on the development of food products from Asteraceae, including *T. lucida*. This is why it is proposed as a promising species to produce harmless functional foods.

7.3.3 *PERSPECTIVES OF* TAGETES LUCIDA *USE IN THE DEVELOPMENT OF FUNCTIONAL FOOD*

One of the main properties of *T. lucida* is its antioxidant activity, which is conferred by the chemical composition it contains, providing a great potential for the development of functional foods. In addition to its antioxidant effect, it has other properties such as antimicrobial, anti-inflammatory, hypoglycemic, and antispasmodic activities [22, 47, 72]. However, the compound giving rise to these activities has not been yet defined. Therefore, it is important to develop research to find if its effects are due to a synergism or antagonism between the compounds. Some compounds containing *T. lucida* as well as some of their properties are shown in Table 7.2.

TABLE 7.2 Properties and Compounds Present in *Tagetes Lucida*

Property	Active Compounds	Reference
Hypoglycemic	Alkaloids	[85]
Antispasmodic		
Antioxidant	Carotenoids, sterols, and polyphenols	[47]
Antifungal	Triterpenes, Polyenes	[28]
Antibacterial	Phenolic compounds, terpenoids	[28]

T. lucida has shown a great antimicrobial spectrum on human, plants, and animals pathogens. *Escherichia coli*, *Proteus mirabilis*, *Klebsiella pneumoniae* and *Salmonella sp.* are some of the bacteria sensible to this plant [28]. Other bacteria sensible to the methanolic extract of this plant are *Escherichia coli*, *Salmonella enteritidis*, *Salmonella typhi*, *Shigella dysenteriae*, and *Shigella flexneri*. On the other hand, it has been demonstrated that *E. coli* has a higher resistance than *S. typhi,* since it shows an inhibition of 33.73% [34].

Additionally, it has been determined that *Candida albicans* and *Staphylococcus aureus* are sensible to the *T. lucida* ethanolic extract, since it showed great inhibition zones, whereas *Diaeretiella rapae* shows smaller inhibition zones [73]. The methanolic extract of *T. lucida* showed an antioxidant effect (IC 50 = 109%) similar to ascorbic acid by the DPPH method [74].

T. lucida had low toxicity on *A. salina* (CL50 = 28 – 454.03 μg/mL) [49, 50, 74] so it is promising for future new food products. Table 7.3 shows the extracted compounds, the solvent used for the extraction and the vegetal part used of *T. lucida.*

TABLE 7.3 Solvent and Part of the Plant to be Used for the Extraction of Compounds

Active compound	Solvent	Vegetable organ	Reference
Estragol, monoterpenes	N/A	Stem, sheet, flower	[44]
3,4-dimethyl ether 7-O-β-D-glucopyranoside and	Methanol	Sheet	[47]
3-(2-O-β-D glucopyranos y l–4-methoxyphenyl)			
7-methoxycoumarin and 6,7dimethoxycoumarin	Hexane	Sheet, Flower, Stem	[28]
Thiophene	Methanol	Root	[73]
Flavonoids, alkaloids and Anthocyanins	Methanol	Stem, sheet, flower	[74]

According to the reviewed information, *T. lucida* and some of its compounds could be used for the development of functional foods. The different phytochemical compounds give rise to different biological activities and provide functionality when incorporated in a food. There is a relationship between the compounds contained in medicinal plants and *T. lucida*, for it has antioxidant properties and therefore plays an important role in cancer prevention.

This review permits to identify the extracts of *T. lucida* that have an antimicrobial effect and could, therefore, be used to preserve food. It should be mentioned that although some compounds of the genus Tagetes are harmful for health and the values of toxicity are low. In this review, it was identified that many of the health benefits can be provided from developed plant foods including the asteracea family, including *T. lucida* and these could be directed to the production of foodstuffs such as juices or solid foods. Some of these properties may be anti-inflammatory, antihyperglycemic, antifungal, and antitumor. These data suggest that *T. lucida* has the potential to formulate functional foods such as infusions, additives, traditional remedy, sauces, and powder to prepare beverages. For this reason, the Asteraceae family offers many favorable prospects for food development.

Although this family has several health benefits, there are still several challenges to overcome, since it contains compounds that show unfavorable effects. For this reason, purification of compounds with biological activity should be done, in order to eliminate harmful compounds. Another challenge in the development of functional herbal foods is the variability of the components; since the quantity and quality of the compounds is affected by the geographic region as well as the season of the year in which the plant is collected.

KEYWORDS

- **functional food**
- **human health**
- ***Tagetes lucida***

REFERENCES

1. Astiasarán, I., & Martínez, A. (1999). Food, composition and properties, 1st Edn.; McGraw-Hill: Spain.
2. Arai, S., (1996). Studies on functional foods in Japan. State of the art. *Biosci. Biotech. Biochem.*
3. Palou, A., & Serra, F., (2000). European Perspectives on Functional Foods. *ANS., 7,* 76–90.
4. Calvo, B., Coral, S., Gómez, C., Royo, C., & Nomdedeu, M. (2011). Nutrition, health and functional foods, 1st Edn.; UNED: Spain.
5. Gupta, D., Dubey, J., & Kumar, M., (2016). Phytochemical analysis and antimicrobial activity of some medicinal plants against selected common human pathogenic microorganisms. *Asian Pac. J. Trop. Dis., 6,* 15–20.
6. *Cos, P., Vlietinck, A. J., Berghe, D. V., & Maes, L., (2006). Anti-infective potential of natural products:* How to develop a stronger in vitro proof-of-concept. *J. Ethno., 106,* 290–302.
7. Sasidharan, S., Chen, Y., Saravanan, D., Sundram, K. M., Latha, L. Y., (2011). Extraction, isolation and characterization of bioactive compounds from plants extracts. *J. Trad. Compl. Altern. Med., 8,* 1–10.
8. Regalado, E. L., Fernandez, M. D., Pino, J. A., Mendiola, J., & Echemendia, O. A., (2011). Chemical composition and biological properties of the leaf essential oil of *Tagetes lucida* Cav. from Cuba. *J. Essent. Oil Res., 23,* 7–63.
9. Bonilla-Jaime, H., Guadarrama-Cruz, G., Alarcon-Aguilar, F. J., Limón-Morales, O., & Vazquez-Palacios, G., (2015). Antidepressant-like activity of *Tagetes lucida* Cav. is mediated by 5-HT(1A) and 5-HT(2A) receptors. *J. Nat. Med., 69,* 463–470.

10. Pal, S., Ghosh, P. K., & Bhattacharjee, P., (2016). Effect of packaging on shelf-life and lutein content of marigold (*Tagetes erecta* L.) flowers. *Recent Pat. Biotechnol., 10*(1), 103–120.
11. Jabeen, A., Mesaik, M. A., Simjee, S. U., Lubna, Bano, S., & Faizi, S., (2016). Anti-TNF-α and anti-arthritic effect of patuletin: A rare flavonoid from *Tagetes patula*. *Int. Immunopharmacol., 36,* 232–240.
12. Parastoo, K., & Gholamreza, K., (2014). Zahra, amirghofran. Anti-oxidative and anti-inflammatory effects of *Tagetes minuta* essential oil in activated macrophages. *Asian Pac. J. Trop. Biomed., 4,* 219–227.
13. López, S. B., López, M. L., Aragón, L. M., Tereschuk, M. L., Slanis, A. C., Feresin, G. E. et al., (2011). Composition and anti-insect activity of essential oils from Tagetes *l.* species (Asteraceae, Helenieae) on *Ceratitis capitata* Wiedemann and *Triatoma infestans* Klug. *J. Agric. Food Chem., 59,* 5286–5292.
14. Yanrui, Xu., Zhiyuan, L., & Seong-Ho, Lee., (2017). Arctigenin represses TGF-b-induced epithelial mesenchymal transition in human lung cancer cells. *Biochem. Biophys. Res. Commun.,* 1–6.
15. Coté, H., Boucher, M. A., Pichette, A., & Legault, J., (2017). Anti-inflammatory, antioxidant, antibiotic, and cytotoxic activities of *Tanacetum vulgare* L. Essential oil and its constituents. *Medicines (Basel), 4,* 34.
16. Ayub, M. A., Hussain, A., Hanif, M. A., Chatha, S., Kamal, G. M., Shahid, M., & Janneh, O., (2017). Variation in phenolic profile, β-carotene and flavonoid contents, biological activities of two tagetes species from Pakistani flora. *Chem. Biodivers., 14,* 1–8.
17. Weiyou, W., Honggao, Xu., Hua, C., Kedong, T., Fuguo, L., & Yanxiang, G., (2016). *In vitro* antioxidant, anti-diabetic and antilipemic potentials of quercetagetin extracted from marigold (*Tagetes erecta* L.) inflorescence residues. *J. Food Sci. Technol., 53,* 2614–2624.
18. Bhattacharyya, S., Datta, S., Mallick, B., Dhar, P., & Ghosh, S., (2010). Lutein content and *in vitro* antioxidant activity of different cultivars of Indian marigold flower (*Tagetes patula* L.) extracts. *J. Agric. Food Chem., 58,* 8259–8264.
19. Ali, N. A., Sharopov, F. S., Al-Kaf, A. G., Hill, G. M., Arnold, N., Al-Sokari, S. S., Setzer, W. N., & Wessjohann, L., (2014). Composition of essential oil from *Tagetes minuta* and its cytotoxic, antioxidant and antimicrobial activities. *Nat. Prod. Commun., 9,* 265–268.
20. Metcalf, R. L., (1982). Insect control technology. In: *Encyclopedia of Chemical Technology* (Vol. 13, pp. 413–485). Wiley-Interscience: New York.
21. Greulach, A. V., & Edison, A. J. *The plants,* 1st Edn.; Limusa: México, 1970.
22. Ciccio, J. F., (2004). A source of almost pure methyl chavicol: Volatile oil from the aerial parts of *Tagetes lucida* (Asteraceae) cultivated in Costa Rica. *Rev. Biol. Trop., 52,* 853–857.
23. Marote, I., Marotti, M., Piccaglia, R., Nastri, A., Grandi, S., & Dinelli, G., (2010). Thiophene occurrence in different *Tagetes* species: Agricultural biomasses as sources of biocidal substances. *J. Food Agric., 90,* 1210–1217.
24. Rzedowsky, J., (1991). The endemism in the Mexican flora: A preliminary analytical appreciation. *Acta Bot. Mex., 15,* 4764.

25. Márquez-Alonso, C., Lara, F., Esquivel, B., & Mata, R. Medicinal Plants of Mexico,1 rd Ed.; UNAM: México, 1999.
26. Aylon, T. T., & Chávez, F. J., Mexico: Natural Resources and Population, 1 st ed.; Limusa: México, 1994.
27. Sarukhán, J., (1995). Biological Diversity, *University of Mexico*, *536*, 3–10.
28. Céspedes, C. L., Ávila, J. G., Martínez, A., Serrato, B., Calderón-Mugica, J. C., & Salgado-Garciglia, R., (2006). Antifungal and antibacterial activities of Mexican tarragon *(Tagetes lucida)*. *J. Agric. Food Chem.*, *54*(10), 853–857.
29. Chasquibol, S., (2003). *Rev. Per. Quím. Ing. Quim.*, *5*(2), 9–20.
30. Astiasarán, I., & Martínez, A. Ecologic and transgenic food. 1 st Ed.; McGraw-Hill: Spain, 1995.
31. Human Medicinal Agents from Plants: Plant derived natural products in drug discovery and development Human medicinal agents from plants; Balandrin, M. F., Kinghorn, A. D., & Farnsworth, N. R.; ACS Symposium Series; American Chemical Society: Washington, DC, 1993.
32. Ávalos, G. A., (2009). Pérez-Urria E. *Serie Fisiología Vegetal.* Reduca (Biología)., *2*(3), 119–145.
33. Roberfroid, M. B., (2000). Concepts and strategy of functional food science: The European perspective. *J. Clin. Nutr.*, *71*, 1664–1669.
34. Kinsella, J. E., Frankel, E., German, B., Kanner, J., & Kinsella, G. J., (1993). Possible mechanisms for the protective role of antioxidants in wine and plant foods. *Food Technol.*, *47*, 85–90.
35. Block, G., Patterson, B., & Subar, A., (1992). Fruit vegetables and cancer prevention: A review of the epidemiological evidence. *Nutr. Cancer*, *18*, 1–29.
36. Burton, G. W., & Lngold, K. V., (1981). Autooxidation of biologica/ molecules. The antioxidant activity of vitamin E and related chainbreaking phenolic antioxidants *in vitro*. *J. Amer. Chem. Soc.*, *103*, 72–77.
37. Abdel-Al, S., & Hucl, P., (1999). A rapid method for quantifiying total anthocyanins in blue aleurone and purple pericarp wheats. *Cereal Chem.*, *76*, 350–354.
38. Kamei, H., Hashimoto, Y., Koide, T., Kojima, T., & Hasegawa, M., (1998). Effect of metanol extracts from red and white wines. *Cancer Biother Radiopharm.*, *13*(6), 447–452.
39. Li, J., Zhang, B., Liu, H., Zhang, X., Shang, X., & Zhao, C., (2016). Triterpenoids from *Ainsliaea yunnanensis* Franch. and their biological activities. *Molecules*, *21*(11), 1481.
40. Caetano, B. F., De Moura, N. A., Almeida, A. P., Dias, M. C., Sivieri, K., & Barbisan, L. F., (2016). Yacon (*Smallanthus sonchifolius*) as a Food Supplement: Health-Promoting Benefits of Fructooligosaccharides. *Nutrients*, *8*(7), 436.
41. Wei-Cai, Z., Wen-Chang, Z., Wen-Hua, Z., Qiang, H., & Bi, S., (2013). The antioxidant activity and active component of *Gnaphalium affine* extract. *Food Chem. Toxicol.*, *58*, 311–317.
42. Pérez-Ortega, G., González-Trujano, M. E., Ángeles-López, G. E., Brindis, F., Vibrans, H., Reyes- Chilpac, R., (2016). *Tagetes lucida* Cav. Ethnobotany, phytochemistry and pharmacology of its tranquilizing properties. *J. Ethnopharm.*, *181*, 221–228.

43. Perossini, M., Guidi, G., Chiellini, S., & Siravo D., (1987). Clinical study on the impexance of mycytic anthocyanidides (Tegens) in the treatment of diabetic and hypertensive retinal microangiopathies, *Ottal Clinical Ocular*., *113*, 1173–1190.
44. Bicchi, C. M., Fresia, P., Rubiolo, D., Monti, C., & Franz, I., (1997). Constituents of *Tagetes lucida* Cav. ssp. lucida essential oil. *J. Flavour Frag*., *12*, 47–52.
45. Stuchlík, M., & Stanislav, Ž., (2002). Vegetable lipids as components of functional foods. *Biomedical* Papers of the Medical Faculty of the University Palacky Olomouc Czechoslovakia, *146*(2), 3–10.
46. Iweriebor, B. C., Okoh, S. O. Nwodo,U. U., Obi, L. C., & Okoh, A. I., (2017). Chemical constituents, antibacterial and antioxidant properties of the essential oil flower of *Tagetes minuta* grown in Cala community Eastern Cape, South Africa. *BMC Complemen Altern Med*., *17*, 351.
47. Aquino, R., Cáceres, A., Morelli, S., & Rastrelli, L., (2002). An extract of *tagetes lucida* and its phenolic constituents as antioxidants. *J. Nat. Prod*., *65*(12), 1773–1776.
48. Cáceres, A., Cano, O., Samayoa, B., & Aguilar, L., (1990). Plants used in guatemala 55 for the treatment of gastrointestinal disorders screening of 84 plants against enterobacteria. *J. Ethnopharm*., *30*, 55–73.
49. Mejia-Barajas, A. et al., (2012). Cytotoxic activity in *Tagetes lucida* Cav. *J. Food Agric*., *24*, 142–147.
50. Mejia-Barajas, J. A., Cajero, J, M., Del Rio, R., & Martínez-Pacheco M., (2011). Scrutiny of the toxic activity of *Tagetes lucida* in *Artemia salina*. *Rev. Latinoamer. Quim*., *38*, 95.
51. Bonilla-Jaime, H., Guadarrama-Cruz, G., Alarcon-Aguilar, F. J., Limón-Morales, O., & Vazquez-Palacios G., (2015). Antidepressant-like activity of *Tagetes lucida* Cav. is mediated by 5-HT1A and 5-HT2A receptors. *J. Nat. Med*., *69*(4), 463–470.
52. Pérez-Ortega G., González-Trujano, M. E., Ángeles-López, G. E., Brindis, F., Vibrans, H., & Reyes- Chilpa R., (2016). *Tagetes lucida* Cav.: Ethnobotany, phytochemistry and pharmacology of its tranquilizing properties. *J. Ethnopharm*., *181*, 221–228.
53. Castañeda, R., Vilela, D., González, M., Mendoza, S., & Escarpa, A., (2013). SU–8/ Pyrex microchip electrophoresis with integrated electrochemical detection for class-selective electrochemical index determination of phenolic compounds in complex samples. *Electrophoresis*., *34*, 2129–2135.
54. Hethelyi, E., Tetenyi, P., Dabi, E., & Danos, B., (1987). The role of mass spectrometry in medicinal plant research. *Biomed Environ Mass Spectrom*., *14*, 627–632.
55. Abdel-Aal, E. S. M., & Rabalski, I., (2015). Composition of lutein ester regio isomers in marigold flower, dietary supplement, and herbal tea. *J. Agric. Food Chem*., *63*(44), 9740–9746.
56. Hadden, W. L., Watkins, R. H., Levy, L. W., Regalado, E., Rivadeneira, D. M., van Breemen, R. B., & Schwartz, S. J., (1999). Carotenoid composition of marigold (*Tagetes erecta*) flower extract used as nutritional supplement. *J. Agric. Food Chem*., *47*(10), 4189–4194.
57. Abdala, L. R., (1999). Flavonoids of the aerial parts from *Tagetes lucida* (Asteraceae). *Biochem. Syst. Ecol*., *27*(7), 753–754.
58. Omer, E. A., Hendawy, S. F., Ismail, R. F., Petretto, G. L., Rourke, J. P., Pintore, G., (2017). Acclimatization study of *Tagetes lucida* L. in Egypt and the chemical characterization of its essential oils. *J. Nat. Prod. Res*., *31*(13), 1509–1517.

59. Cicció, J. F., (2004). A source of almost pure methyl chavicol: Volatile oil from the aerial parts of *Tagetes lucida* (Asteraceae) cultivated in Costa Rica. *Rev. Biol. Trop.*, *52*(4), 853–857.
60. Armas, K., Rojas, J., Rojas, L., & Morales, A., (2012). Comparative study of the chemical composition of essential oils of five Tagetes species collected in Venezuela. *Nat. Prod. Commun.*, *7*(9), 1225–1226.
61. Ezzat, S. M., Salama, M. M., Mahrous, E. A., & Abdel-Sattar, E., (2017). Antiprotozoal activity of major constituents from the bioactive fraction of *Verbesina encelioides*. *Nat. Prod. Res*. *31*(6), 676–680.
62. Inafuku, M., Takata, K., Taira, N., Nugara, R., Kamiyama, Y., & Oku, H., (2016). Monogalactosyldiacylglycerol: An abundant galactosyllipid of *Cirsium brevicaule* A. GRAY leaves inhibits the expression of gene encoding fatty acid synthase *Phytomedicine.*, *13*(5).
63. Pandey, A. K., & Singh, P., (2017). The genus Artemisia: A 2012–2017 review on chemical composition, antimicrobial, insecticidal and antioxidant activities of essential oils. *Medicines (Basel)*, *4*, 68.
64. Coté, H., Boucher, M. A., Pichette, A., & Legault, J., (2017). Anti-inflammatory, antioxidant, antibiotic, and cytotoxic activities of *Tanacetum vulgare* L. Essential oil and its constituents. *Medicines (Basel)*, *4*, 34.
65. Lehbili, M., Alabdul, M. A., Kabouche, A., Voutquenne-Nazabadioko, L., Abedini, A., Morjani, H., Sarazin, T., Gangloff, S. C., Kabouche, Z.(2017). Oleanane-type triterpene saponins from *Calendula stellata. Phytochemistry.*, *144*, 33–42.
66. Talawar, S., Harohally, N. V., Ramakrishna, C., & Suresh, K. G., (2017). Development of wheat bran oil concentrates rich in bioactives with antioxidant and hypolipidemic properties. *J. Agric. Food Chem.*, *65*(45), 9838–9848.
67. Chan, K., (2003). Some aspects of toxic contaminants in herbal medicines. *Chemosphere*, *52*(9), 1361–1371.
68. Pak, E., Esrason, K. T., & Wu, V. H., (2004). Hepatotoxicity of herbal remedies: An emerging dilemma. *Prog. Transplant.*, *14*(2), 91–96.
69. Talawar, S., Harohally, N. V., Ramakrishna, C., & Suresh, Kumar, G., (2017). Development of wheat bran oil concentrates rich in bioactives with antioxidant and hypolipidemic properties. *J. Agric. Food Chem.*, 1–38.
70. Sun, Z., Zhang, K., Chen, C., Wu, Y., Tang, Y., Georgiev, M., Zhang, X., Lin, M., & Zhou, M., (2017). Biosynthesis and regulation of cyanogenic glycoside production in forage plants. *Appl. Microbiol. Biotechnol. 102*(1), 9–16.
71. Silveira, R., Megías, M., & Molina B., (2003). Functional foods and optimal nutrition, near or far?. *Rev. Esp. Salud. Pública.*, *77*(3), 317–331.
72. Marriott, B. M., (2000). Functional foods an ecologic perspective. *Am. J. Clin. Nutr.*, *71*(6 Suppl):1728S–34S.
73. Elsayed, A. O., Saber, F. H., Nour El-Deen, A. M., Zaki, F. N., Abd-Elgawad, M. M., Kandeel, A. M., Ibrahim, A. K., & Ismail, R. F., (2015). Some biological activities of *Tagetes lucida* plant cultivated in Egypt. *Adv. Environ. Biol.*, *9*(2), 82–88. Available from: https://www.researchgate.net/publication/273814244_Some_Biological_Activities_of_Tagetes_lucida_Plant_Cultivated_in_Egypt [accessed Aug 01 2018].

74. Téllez-López, M. A., Treviño-Neávez, J. F., Verde-Star, M. J., Rivas-Morales, C., Oranday-Cárdenas, A., Moran-Martínez, J., Serrano-Gallardo, L. B. & Morales-Rubio M. E. (2013). Evaluation of the effect of methanol extract of *Tagetes lucida* on testicular function and sperm quality in male Wistar rats. *Rev. Mex. Cienc. Farmac.*, *44*(4), 43–52.
75. Karakose, H., Müller, A., & Kuhnert, N., (2015). Profiling and Quantification of Phenolics in *Stevia rebaudiana* Leaves. *J. Agric. Food Chem.*, *63*(41), 9188–9198.
76. Gylling, H., & Miettinen T. A., (2005). The effect of plant stanol and sterol enriched foods on lipid metabolism, serum lipids and coronary heart disease. *Ann. Clin. Biochem.*, *42*, 254–253.
77. Bouic, P. J., (2001). The role of phytosterols and phytosterolins in immune modulation: A review of the past 10 years. *Curr. Opin. Clin. Nutr. Metab. Care.*, *4*(6), 471–475.
78. Yeh, C. T., Yen, G. C., (2003). Effects of phenolic acids on human phenolsulfotranferases in relation to their antioxidant activity. *J. Agric. Food Chem.*, *51*, 1474–1479.
79. Nohynek, L., Meier, C., Kähkönen, M., Heinonen, M., Hopia, A., Oksman-Caldentey, K. M., (2001). Antimicrobial properties of phenolic compounds from berries. *J. Appl. Microbiol.*, *90*(4), 494–507.
80. Rivas Gonzalo, J. C., & García Alonso M., (2002). Flavonoids in plant foods: Structure and antioxidant activity, *Alim. Nutri. Salud.*, *9*, 31–38.
81. Hertog, M. G. L., Kromhout, D., Aravanis, C., Blackburn, H., Buzina, R., & Fidanza, F., (1995). Flavonoid intake and long-term and risk of coronary heart disease and cancer risk in the Seven Countries study. *Arch. Intern. Med.*, *155*(4), 381–386.
82. Hoult, J. R., & Payá, M., (1996). Pharmacological and biochemical actions of simple coumarins. Natural products with therapeutic potential. *Gen. Pharmacol-Vasc. S.*, *27*, 713–722.
83. Castaneda-Ovando, A., Pacheco-Hernández, L., Paez-Hernández, E., Rodríguez, J. A., & Galán-Vidal, C. A., (2009). Chemical studies of anthocyanins: A review. *Food Chem.*, *113*, 859–871.
84. Escribano-Bailon, M., Beulga-Santos, C., & Rivas-Gonzalo, J. C., (2004). Anthocyanins in Cereals. *J. Chromatogr.*, *1054*, 129–141.
85. Ibarra, M. J., Cantú, P. C., Verde, M. J. & Oranday, A., (2009). Phytochemical characterization and hypoglycemic effect of the *Tecoma stans* and its relationship to the presence of chromium as a factor for glucose tolerance. Inf. Tecnol., *20*(5), 55–65.
86. Aponte, M., Calderón, M., Delgado, A., Herrera, I., Jiménez, Y., Ramírez, Z., Rojas, J., & Toro, Y., (2008). Phytochemicals. INN. *Dirección de Investigaciones Nutricionales*, Caracas, Venezuela.

CHAPTER 8

ANTIDIABETIC PROPERTIES OF *PETROSELINUM CRISPUM*

MÓNICA NAYELI VALENCIA LÓPEZ,
CRYSTEL ALEYVICK SIERRA RIVERA, and
SONIA YESENIA SILVA BELMARES

*Food Research Department, Faculty of Chemistry,
Autonomous University of Coahuila, Blvd. Venustiano Carranza,
Colony Republic, Zip code 25280, Saltillo, Coahuila, Mexico,
E-mail: yesenia_silva@uadec.edu.mx*

ABSTRACT

Diabetes Mellitus stands for a group of multiple metabolic disorders in which the insulin secretion is insufficient. The effectiveness and safety of the medication are still challenges, inasmuch as the commonly used drugs cause known side effects. Some phytochemicals behave as active principles that promote a hypoglycemic action. This study presents research data regarding the phytochemical properties of *Petroselinum crispum* and its potential use in nutraceutical products such as supplements, drinks or enriched foods.

8.1 INTRODUCTION

Diabetes mellitus is a chronic metabolic disease in which there is an increase in the blood glucose level due to the deficiency in insulin secretion of the body [1]. This disease can trigger secondary disorders such as neuropathies, retinopathies, renal failure and skin complications [2]. Additionally, it is known that 25% of the world population is affected by this disease [3], which means that more than 347 million people suffer from it [4]. The global healthcare expenditures have been estimated to be of US$ 376 billion worldwide in 2010 and US$490 billion by the year 2030 [5]

The traditional pharmacologic treatments usually consist of hypoglycemic agents and insulin, these therapies, however, cause severe side effects when used chronically [2]. This is why alternative or complementary treatments are being studied.

For many centuries, people have used plants as empirical treatments. Almost 800 plants are currently used as folk remedies to control diabetes mellitus such as *Petroselinum* crispum [6]. The present review compiles information on the properties of *Petroselinum crispum* as a preventive and corrective treatment for this disease.

8.2 TYPE 2 DIABETES MELLITUS AND THE ROLE OF ANTIOXIDANTS

Type 2 diabetes is formed by a group of metabolic disorders identified by three characteristics: an increased liver glucose production, insulin resistance and insufficient insulin secretion [7]. The first stage of the illness is an insulin resistance that is compensated by an excessive secretion of insulin by the β-cells. The second stage is the insufficient insulin production by a deficient pancreatic functional reserve which leads to stage three, an increment of blood glucose concentration due to an irreversible β-cells damage [8].

Once the patient gets to stage three, the illness is irreversible. Nevertheless, by protecting the cells or reducing insulin resistance we could reduce the risk of the disease up to 36% [9]. In diabetes, the excess glucose increases the formation of reactive oxygen species, causing cellular damage and eventually destruction of pancreatic β-cells. Antioxidants have shown the capacity to prevent this destruction [10], and also the capacity to inhibit the enzymes α-amylase and α-glucosidase, modulating postprandial glycaemia and controlling starch metabolism [11]. When the enzymes α-glucosidase and α-amylase are blocked, the processes of digestion and absorption of glucose are obstructed [12].

8.3 α-AMYLASE AND α-GLUCOSIDASE INHIBITORY POTENTIAL

An efficient way to balance the glycemic levels is through the regulation or inhibition of carbohydrate-hydrolyzing enzymes [13] α-amylase is one of the most important enzymes secreted by the pancreas and salivary glands

[14]. The digestion of carbohydrates begins in the mouth by its breakage into smaller oligomers by the salivary α-amylase. Subsequently, the oligomers are hydrolyzed by the pancreatic α-amylase and released into the intestinal lumen, [15] decreasing post-prandial hyperglycemia [16].

α-glucosidases are glycosylphosphatidyl anchored enzymes positioned on the surface of the intestinal cells [17]. Their function is to catalyze the last step of the digestion of carbohydrates [18] and as a result, they also reduce post-prandial hyperglycemia.

Given the importance that these therapeutic mechanisms represent, there are drugs who inhibit these enzymes, such as acarbose, miglitol, and voglibose [19] classified as third category oral hypoglycemic agents. However, the standard dosage usually inhibits an excessive amount of pancreatic enzyme, leading to abnormal bacterial fermentation in colon [11] causing serious gastrointestinal side effects [12].

8.4 ANTIDIABETIC PHYTOCHEMICALS

The identification of α-glucosidase and α-amylase inhibitors isolated from plants could lead to the development of new treatment options without digestive side effects [20]. Many studies have demonstrated that because of their larger molecular weight, phytochemicals easily blind to enzymes and inhibit their activity with tremendous potency [21]. The main groups of compounds that demonstrate hypoglycemic activity are described below.

8.4.1 TERPENES

Terpenes are the primary component of vegetal essential oils. Isoprene units form the skeleton of the more of 30,000 different terpenes known at present (isoprene rule) [22]. Many researches have shown the qualities of this compound group in the battle against diabetes through the inhibition of the enzyme α-amylase [23].

8.4.2 FLAVONOIDS

Flavonoids are part of the polyphenolic family. They have a 15-carbon skeleton with two phenyl rings and a heterocyclic ring, being responsible for the pigments present in most flowers and leaves [24]. Many studies have reported numerous benefits on human health when consumed regularly [25]. In a study done on diabetic rats, isorhamnetin decreased notably serum glucose and sorbitol concentrations [26]. Additionally, kaempferol enhanced the expression of the proteins AKT and Bcl-2, the signaling of cAMP and improved the synthesis of insulin [27]. It also stimulated the synthesis of new glucose transporters [28] and reduced serum HbA1c levels enhancing insulin resistance [29]. The mechanism of action of the flavonoids includes antioxidant reactions, gut transport alterations, central nervous system changes and increased insulin sensitivity [30]. They are also involved in the reduction of cholesterol and fat reducing coronary heart disease, linked to metabolic syndrome [31].

8.4.3 COUMARINS

Coumarins are aromatic benzopyrene chemical compounds known by their anticoagulant and anti-edema properties [32]. In diabetic rats, they have shown a decrease of plasma glucose, HbA1c and an increase of insulin, exposing protective effects on glycoproteins component [33]. Coumarins improve pancreatic function stimulating insulin through anti-inflammatory, anti-oxidative, and anti-apoptotic activity [34]. It decreases gluconeogenesis through the increment of hexokinase activity and glucose–6-phosphate dehydrogenase. Coumarins also increase glucose decomposition and reduce the enzymes glucose 6-phosphatase and fructose–1, 6-biss phosphatase [35].

8.5 HYPOGLYCEMIC PLANTS FROM THE *UMBELLIFERAE* FAMILY

The *Umbelliferae* family has more than 3700 species, including famous and economically important herbs such as *Coriandrum sativum, Angelica gigas, Angelica sinesis, Angelica acutiloba, Centella asiatica,* among others. Many of these plants have shown antidiabetic properties, such as *Angelica acutiloba,* cultivated mainly in Asia. It improves insulin sensitivity ameliorating the disturbance in hepatic glucose [14]. *Ligusticum porter*, a Mexican plant, has revealed an α-glucosidase inhibition, diminishing glucose absorption and also stimulating insulin secretion in high doses [36]. In recent studies, *Cnidium monnieri* (used in folk Chinese medicine) improved insulin resistance in a high-fat and high-sucrose diet through the increment of adiponectin release by the PPARα/γ pathway [37]. *Cuminum cyminum* controls oxidative stress and inhibits advanced glycated end products, controlling diabetes significantly [38]. *Angelica hirsutiflora* is capable of elevating insulin secretion, acting as a secretagogue regulating prandial glucose in non-insuline-dependent patients [39]. *Arracacia tolucensis*, a plant from the north of Mexico, decreases blood levels significantly through the action of pyranocoumarin controlling body weight and decreasing aspartate aminotransferase and alanine aminotransferase, inhibiting cytokines expression [40]. Rutin in *Eryngium bornmuelleri* modulates the metabolism of sugars and fats inhibiting relevant digestive enzymes [41]. *Ducrosia anethifolia Boiss.* is used as a flavoring additive in Europe and the Middle East, this plant inhibits the enzymes α-amylase, α-glucosidase and α-galactosidase, ameliorating the antioxidant markers and the glucolytic and gluconeogenic enzymes. All these effects improve kidney metabolism and reduces blood glucose concentrations [42]. *Centella asiatica* is used as a traditional remedy in Africa and India. Chronic consumption of this plant reduces serum LDL and cholesterol, obstructing the absorption of intestinal saccharidase enzymes and glucose-fiber binding [10]. *Cumin cyminum L.* helps to preserve the integrity of β-cells of the pancreatic islets through the action of prenylflavonoids, due to the homeostasis on glucose and lipid concentrations that it creates, having the same effect of Orlistat on weight loss and insulin metabolism [43]. *Angelica sinensis,* a famous Chinese plant, possesses a variety of pharmacological properties in diabetes mellitus. It decreases body weight in prediabetic and diabetic individuals, increasing hepatic and muscle glycogen and improving insulin resistance [44]. *Angelica decursiva* is a plant widely distributed in China, Japan, and Korea. The extracts of this plant have been associated to the inhibition of α-glucosidase, PTP1B, and ONOO-mediated tyrosine nitration. Postprandial hyperglycemia generates nitrotyrosine making it a diabetes risk factor. Therefore, this plant is a promising treatment for diabetes and metabolic syndromes [45]. On the other hand, kidney, and liver are in danger of suffering pathological changes in diabetes, *Aegopodium podagraria L.* has nephroprotective and hepatoprotective actions, being a potential supplement to regular hypoglycemic treatments [46]. *Angelica gigas* is a biennial plant commonly found in Japan, China, and Korea. It has the potential to improve insulin sensibility through the activation of the AMPK pathway [47]. *Heracleum dissectum L.* is a plant with a high content of coumarins that has the capacity to improve insulin resistance and increase glucose uptake used by peripheral tissues [48]. *Sphallerocarpus gracilis* protects pancreatic β-cells from free radical damage, reducing apoptosis of pancreatic β-cells and stimulating insulin secretion [49]. *Coriandrum sativum L.* normalizes glycemic levels, decreases elevated levels of insulin, IR, TC, LDL-cholesterol, and TG [50].

8.6 HYPOGLYCEMIC PROPERTIES OF *PETROSELINUM CRISPUM*

Petroselinum crispum from the *Umbelliferae* family, is native to the Mediterranean because of its growth needs [6]. Table 8.1 is presented with the taxonomy of the plant and Table 8.2 with the cultivation conditions. The Mediterranean population believed it protected food from external contamination [51]. On the other hand, Romans alleged that the strength of the gladiators increased by the consumption of the plant on a regular basis [52]. Additionally, Iranian medicine claimed it to be antimicrobial, antiseptic, diuretic, and hypoglycemic [6].

Petroselinum crispum contains a rich presence of flavonoids, carotenoids, coumarins, and other compounds. Table 8.3 shows its main phytochemical compounds [53].

8.6.1 IN VIVO *MODELS*

In order to prevent toxicity and to understand the hypoglycemic mechanisms of the plant, *in vivo* models have been used to mimic the structural and functional characteristics of humans [63]. So far, the only *in vivo* models reported are on Swiss Albino rats. The researches were carried out using diabetic animals induced by streptozotosin as described in Table 8.4.

The studies used 6-months-old male Swiss Albino rats, administrating a daily dose of 2g/kg of an aqueous extract of leaves for 28–45 days. All studies showed hypoglycemic activity. Some of the most outstanding results were the decrease of the aorta lipid peroxidation, an increment in SOD activity and a decrease on the liver LPO levels [64].

TABLE 8.1 Taxonomy of *Petroselinum Crispum*

Category	Description
Kingdom	*Plantae*
Subkingdom	*Viridiplantae*
Infrakingdom	*Streptophyta*
Superdivision	*Embryophyta*
Division	*Tracheophyta*
Subdivision	*Spermatophytina*
Class	*Magnoliopsida*
Superorder	*Asteranae*
Order	*Apiales*
Family	*Umbelliferae*
Genus	*Petroselinum*
Species	*Crispum* [54]

TABLE 8.2 Cultivation Conditions of *Petroselinum Crispum*

Environment	Conditions
Soil	Moist, well drained and loamy
Sun exposure	Full-sun and part-sun environments
Temperature	It grows best between 20–32°C [55]
Water	Drought-tolerant
Spacing	25 cm apart [56]

TABLE 8.3 Main Phytochemical Compounds Identified in Petroselinum Crispum

Compound group	Chemical Category	Chemical Compound
Terpenes	Carotenoids	β-Carotene
		Luiteine
	Sesquiterpene [57]	Violaxanthin
		Neoxanthin [53]
		Crispane
		Crispanone
Flavonoids [58, 59]	Flavone	Luteolin
	Flavonol	Apigenin
		Isorhamnetin [60]
		Kaempferol [61]
Coumarins	Furanocoumarins O-Methylated	8-Methoxypsoralen
	Furanocoumarins Aglycones	5-Methoxypsoralen
	Meroterpene furanocoumarin ether [62]	Imperatorin

8.6.2 IN VITRO *MODELS*

The *in vitro* techniques are procedures taken place in controlled environments outside of a living organism [69]. Some of the *in vitro* studies of *P. crispum* include the evaluation of bone oxidative stress and the antioxidant enzymes. The plant extracts induced a significant increase of the glutathione-S-transferase and glutathione peroxidase activities [70]. On the other hand, in gastric samples treated with the plant showed that

TABLE 8.4 Properties of *Petroselinum Crispum*: In Vivo Models

Type of rats	Doses	Extract	Methodology	Results
6-month-old male Swiss Albino rats weighing 150–200g	2 g/kg daily for 28 days	Leaves Aqueous extract	LPO levels in liver homogenates were estimated by Led-wozyw's method	The liver LPO levels decreased when compared to the untreated diabetic rats [53]
6-month-old male Swiss Albino rats	2 g/kg daily for 28 days	Leaves Aqueous extract	Obtainment of pancreatic tissue samples after an 8-hour fast.	No difference found between the amount of insulin in B cells in extract treated and untreated diabetic rats [64]
6-month-old male Swiss Albino rats weighing 150–200 g	2 g/kg daily for 28 days	Leaves Aqueous extract	The aorta and heart were taken from each killed rat. As an index of lipid peroxidation in aorta and heart tissue homogenates the malondialdehyde (MDA) contents were determined by the spectrophotometric method	The aorta lipid peroxidation was found to be decreased significantly compared to the diabetic animals [65]
6-month-old male Swiss Albino rats weighing 150–200 g	2 g/kg daily for 28 days	Leaves Aqueous extract	Removal of the dermis of the killed animals and Skin total protein levels were measured by the method of Lowry	No difference was found [66]
Albino rats weighing 160–130 g	2 g/kg daily for 45 days	Leaves Aqueous extract	Blood samples were taken to determine the SOD through the hemolysis of red blood cells. Removal of the pancreas of the killed animals	Increment in SOD activity on diabetic rats, a significant decrease in percentage area of caspase-3 expression in pancreatic tissues [67]
6-month-old male Swiss Albino rats	2 g/kg daily for 28 days	Leaves Aqueous extract	Removal of ocular lenses of the killed animals, determination of total protein, protein glycation and glutathione levels of the lens homogenates	No significant decrease found [68]

the histopathologic damage in mucus layer was drastically reduced [71]. Another model employed to evaluate the properties of *P. crispum* is protein spectrophotometric determination by bicinchoninic acid technique. This test is used to evaluate the anti-hyperuricemia effect. The plant proved to be capable of reducing uric acid levels [72]. Additionally, it has been proved *P. crispum* has compounds that have relaxing properties to smooth ileum muscle [73].

8.6.3 QUALITATIVE AND QUANTITATIVE PHYTOCHEMICAL CHARACTERIZATION

Qualitative trials identify the presence of specific metabolites and are used as preliminary tests. Quantitative assays, on the other hand, are used to evaluate specific content. An example is the determination of total phenolic content. This experiment was performed using the Folin-Ciocalteau reagent. The results for the absorbance of *P. crispum* measured at a wavelength of 725 nm [74].

Another quantitative test is the determination of the ferric reducing activity, in which the absorbance of the treated sample is measured at a wavelength of 595 nm. This examination was used to evaluate the iron-reducing activity of *P. crispum* extracts. The results revealed that the plant provides antioxidant protection [75].

One more quantitative assay is the determination of antioxidant capacity using glutathione. This analysis measured the antioxidant capacity of *P. crispum* extract. The results showed that the percentage of oxidized glutathione decreased significantly [76].

8.6.4 NUTRACEUTICAL APPLICATIONS

The plants of the *Umbelliferae* family have innumerable biological properties that can be incorporated in the development of nutraceutical or nutritional supplements given its health benefits. An example is the development of new products such as supplemented pasta, a food widely consumed by the world population. In this research, part of the wheat flour was replaced with powdered leaves of the plant. The fortification of the pasta increased the phenolic levels and thus the antioxidant activity [77].

Also, the incorporation of the same plant to cookies slowed down the process of lipid oxidation and increased the antioxidant capacity of the product [78]. Other members of the *Umbelliferae* family are also used as everyday nutritional supplements like beverages and various formulations of spices. These plants are hypocaloric, rich in oils and other secondary metabolites [79].

8.7 CONCLUDING REMARKS AND FUTURE PERSPECTIVE

For many centuries, the *Umbelliferae* family plants have been used in folk medicine for the management of many illnesses such as hypoglycemia. Diabetes mellitus is a metabolic chronic disease in which there is an increment of glucose due to the deficiency in insulin secretion triggering secondary pathologies in the long run through the increment of levels of reactive oxygen species produced by the excess glucose. *Petroselinum crispum* has a rich content of terpenes, flavonoids, and coumarines and has proved to have hypoglycemic properties that could lead to the prevention or improvement of the disease. The action mechanisms are still not completely identified but it is believed that it has the capacity to inhibit the carbohydrate hydrolyzing enzymes α-amylase and α-glucosidase. Other assays showed its capacity to decrease aorta lipid peroxidation, increment SOD activity and decrease liver LPO levels. There are some nutraceutical applications through supplementation of pasta and cookies that showed increased postprandial phenolic levels and antioxidant activity. This review demonstrates the importance of secondary metabolites from the *Umbelliferae* family, in particular *Petroselinum crispum,* inasmuch as it has the potential to work as a complementary treatment for diabetes mellitus and other metabolic illnesses. The little information available on the subject proves that it is an excellent source for nutraceutical products such as supplements, drinks or enriched food to be exploited in the future.

8.8 CONFLICT OF INTEREST STATEMENT

We declare that we have no conflict of interest.

KEYWORDS

- **bioactive compounds**
- **diabetes**
- **hypoglycemia**
- **phytochemicals**

REFERENCES

1. Kang, S. J., Park, J. H. Y., Choi, H. N., & Kim, J. I., (2015). A-glucosidase inhibitory activities of myricetin in animal models of diabetes mellitus. *Food Sci. Biotechnol.*, *24*(5), 1897–1900.
2. Bahmani, M., Zargaran, A., Rafieian-Kopaei, M., & Saki, K., (2014). Ethnobotanical study of medicinal plants used in the management of diabetes mellitus in the uremia, Northwest Iran. *Asian Pac. J. Trop. Med.*, *7*(1), 348–354.
3. Arumugam, G., Manjula, P., & Paari, N., (2013). A review: Anti diabetic medicinal plants used for diabetes mellitus. *J. Acute Dis.*, *2*(3), 196–200.
4. Meza, R., Barrientos-Gutierrez, T., Rojas-Martinez, R., Reynoso-Noverón, N., Palacio-Mejia, L. S., Lazcano-Ponce, E. et al., (2015). Burden of type 2 diabetes in Mexico: Past, current and future prevalence and incidence rates. *Prev. Med. (Baltim).*, *8*, 445–450.
5. Zhang, P., Zhang, X., Brown, J., Vistisen, D., Sicree, R., Shaw, J., & Nichols, G., (2010). Global healthcare expenditure on diabetes for 2010 and 2030. *Diabetes Res. Clin. Pract.*, *87*(3), 293–301.
6. Farzaei, M. H., Abbasabadi, Z., Reza, M., Ardekani, S., & Rahimi, R., (2013). Parsley : A review of ethnopharmacology, phytochemistry and bio- logical activities. *J. Tradit. Chinese Med.*, *33*(6), 815–826.
7. Rina, R. G., & Hormis, A. P., (2014). Perioperative management of diabetes mellitus and corticosteroid insufficiency. Surgery (Oxford)., *32*(10), 558–562.
8. Zaccardi, F., Webb, D. R., Yates, T., & Davies, M. J., (2016). Pathophysiology of type 1 and type 2 diabetes mellitus: A 90-year perspective. *Postgrad. Med. J.*, *92*(1084), 63–69.
9. Chiasson, J. L., Josse, R. G., Gomis, R., Hanefeld, M., Karasik, A., & Laakso, M., (2002). Acarbose for prevention of type 2 diabetes mellitus: The STOP-NIDDM randomised trial. *Lancet*, *359*(9323), 2072–2077.
10. Sabiu, S., Neill, F. H. O., & Ashafa, A. O. T., (2016). Kinetics of α-Amylase and α-glucosidase inhibitory potential of *Zea Mays Linnaeus (Poaceae)*, *Stigma Maydis* aqueous extract: An *in vitro* assessment. *J. Ethnopharmacol. 183*, 1–8.
11. Duraiswamy, A., Shanmugasundaram, D., Sheela, C., Cherian, S. M., & Mammen, K., (2015). Development of an antidiabetic formulation (ADJ6) and its inhibitory activity against α-amylase and α-glucosidase. *J. Tradit. Chinese Med. Sci.*, 10–14.

12. Yin, Z., Zhang, W., Feng, F., Zhang, Y., & Kang, W., (2015). α-Glucosidase inhibitors isolated from medicinal plants. *Food Sci. Hum. Wellness*, *3*(3–4), 136–174.
13. Proença, C., Freitas, M., Ribeiro, D., Oliveira, E. F. T., Sousa, J. L. C., Tomé, S. M., Ramos, M. J., Silva, A. M. S., Fernandes, P. A., & Fernandes, E., (2017). α-Glucosidase inhibition by flavonoids : An *in vitro* and *in silico* structure activity relationship study. *J. Enzyme Inhib. Med. Chem.*, *32*(1), 1216–1228.
14. De Sales, P. M., De Souza, P. M., Simeoni, L. A., Magalhães, P. De O., & Silveira, D., (2012). α-Amylase inhibitors. A review of raw material and isolated compounds from plant source. *J. Pharm. Pharm. Sci.*, *15*(1), 141–183.
15. Brayer, G. D., Luo, Y., & Withers, S. G., (1995). The structure of human pancreatic α-amylase at 1.8. A resolution and comparisons with related enzymes. *Protein Sci.*, *4*(9), 1730–1742.
16. Loizzo, M. R., Marrelli, M., Pugliese A., Conforti, F., Nadjafi, F., Menichini, F., Tundis, R., (2016). *Crocus cancellatus* subsp. *Damascenus* stigmas: Chemical profile, and inhibition of α-amylase, α-glucosidase and lipase, key enzymes related to type 2 diabetes and obesity. *J. Enzyme Inhib. Med. Chem.*, *31*(2), 212–218.
17. Sivasothy, Y., Kong, L., Kok, L., Litaudon, M., & Awang, K., (2015). A potent α-glucosidase inhibitor from *Myristica cinnamomea* king. *Phytochemistry.*, 122(2016), 254–269.
18. Chu, Y., Wu, S., & Hsieh, J., (2014). Isolation and characterization of α-glucosidase inhibitory constituents from *Rhodiola crenulata. FRIN*, *57*, 8–14.
19. Hakamata, W., Kurihara, M., Okuda, H., Nishio, T., & Oku, T., (2009). Design and screening strategies for α-glucosidase inhibitors based on enzymological information, 3–12.
20. Mata, R., Cristians, S., & Escando, S., (2012). Mexican antidiabetic herbs: Valuable sources of inhibitors of α-glucosidases. *J. Nat. Prod.76* (3), 468–483.
21. Barrett, A. H., Farhadi, N. F., & Smith, T. J., (2018). Slowing starch digestion and inhibiting digestive enzyme activity using plant flavanols/tannins—A review of efficacy and mechanisms. *LWT – Food Sci. Technol. 87*, 394–399.
22. Martin, D. M., (2003). Induction of volatile terpene biosynthesis and diurnal emission by methyl jasmonate in foliage of Norway spruce. *Plant Physiol.*, *132*(3), 1586–1599.
23. Uddin, N., Hasan, M. R., Hossain, M. M., Sarker, A., Hasan, A. H. M. N., Islam, A. F. M. M., Chowdhury, M. M. H. & Rana, M. S., (2014). *In vitro* α-amylase inhibitory activity and *in vivo* hypoglycemic effect of methanol extract of *citrus macroptera montr. Fruit. Asian Pac. J. Trop. Biomed.*, *4*(6), 473–479.
24. Hossain, M. K. Dayem, A. A., Han, J. Yin, Y., Kim, K., Saha, S. K. Yang, G. M., Choi, H. Y., & Cho, S. G., (2016). Molecular mechanisms of the anti-obesity and antidiabetic properties of flavonoids. *Int. J. Mol. Sci.*, *17*(4), 569.
25. Cook, N. C., & Samman, S., (1996). Flavonoids—chemistry, metabolism, cardioprotective effects, and dietary sources. *J. Nutr. Biochem.*, *7*(2), 66–76.
26. Yokozawa, T., Kim, H. Y., Cho, E. J., Choi, J. S., & Chung, H. Y., (2002). Antioxidant effects of isorhamnetin 3,7-Di-O-β-D-glucopyranoside isolated from mustard leaf (*Brassica Juncea*) in rats with streptozotocin-induced diabetes. *J. Agric. Food Chem.*, *50*(19), 5490–5495.

27. Zhang, Z., Ding, Y., Dai, X., Wang, J., & Li, Y., (2011). Epigallocatechin–3-gallate protects pro-inflammatory cytokine induced injuries in insulin-producing cells through the mitochondrial pathway. *Eur. J. Pharmacol.*, *670*(1), 311–316.
28. Folador, P., Figueiredo, M. S. R. B., Pizzolatti, M. G., Leite, L. D., Zanatta, L., & Silva, F. R. M. B., (2008). Insulinomimetic effect of kaempferol 3-neohesperidoside on the rat soleus muscle. 532–535.
29. Zhang, Y., & Liu, D., (2011). Flavonol kaempferol improves chronic hyperglycemia-impaired pancreatic beta-cell viability and insulin secretory function. *Eur. J. Pharmacol.*, *670*(1), 325–332.
30. Ota, A., & Ulrih, N. P., (2017). An overview of herbal products and secondary metabolites used for management of type two diabetes. *Front. Pharmacol.*, *8*, 1–14.
31. Lay, M. M., Karsani, S. A., Mohajer, S., & Abd Malek, S. N., (2014). Phytochemical constituents, nutritional values, phenolics, flavonols, flavonoids, antioxidant and cytotoxicity studies on *Phaleria Macrocarpa* (Scheff.) boerl fruits. *BMC Complement. Altern. Med.*, *14*(1), 152.
32. Rajarajeswari, N., & Pari, L., (2011). Antioxidant role of coumarin on streptozotocin – Nicotinamide-induced type 2 diabetic rats., *25*(6), 355–361.
33. Vinayagam, R., Xiao, J., & Xu, B., (2017). An insight into anti-diabetic properties of dietary phytochemicals. *Phytochem. Rev.* *16*(3), 535–553.
34. Li, H., Yao, Y., & Li, L., (2017). Coumarins as potential antidiabetic agents. *J. Pharm. Pharmacol.*, *69*(10), 1253–1264.
35. Bahmani, M., Golshahi, H., Saki, K., Rafieian-Kopaei, M., Delfan, B., & Mohammadi, T., (2014). Medicinal plants and secondary metabolites for diabetes mellitus control. *Asian Pacific J. Trop. Dis.*, *4*(2), 687–692.
36. Brindis, F., Rodríguez, R., Bye, R., González-Andrade, M., & Mata, R., (2011). (Z)–3-Butylidenephthalide from *Ligusticum Porteri*, an α-glucosidase inhibitor. *J. Nat. Prod.*, *74*(3), 314–320.
37. Qi, Z., Xue, J., Zhang, Y., Wang, H., & Xie, M., (2011). Osthole ameliorates insulin resistance by increment of adiponectin release in high-fat and high-sucrose-induced fatty liver rats. *Planta Med.*, *77*(3), 231–235.
38. Jagtap, A. G., & Patil, P. B., (2010). Antihyperglycemic activity and inhibition of advanced glycation end product formation by *cuminum cyminum* in streptozotocin induced diabetic rats. *Food Chem. Toxicol.*, *48*(8–9), 2030–2036.
39. Leu, Y. L., Chen, Y. W., Yang, C. Y., Huang, C. F., Lin, G. H., Tsai, K. S., Yang, R. S., & Liu, S. H., (2009). Extract isolated from *angelica hirsutiflora* with insulin secretagogue activity. *J. Ethnopharmacol.*, *123*(2), 208–212.
40. García-Galicia, M. C., Burgueño-Tapia, E., Romero-Rojas, A., García-Zebadúa, J. C., Cornejo-Garrido, J., & Ordaz-Pichardo, C., (2014). Anti-hyperglycemic effect, inhibition of inflammatory cytokines expression, and histopathology profile in streptozotocin-induced diabetic rats treated with *arracacia tolucensis* aerial-parts extracts. *J. Ethnopharmacol.*, *152*(1), 91–98.
41. Dalar, A., Türker, M., Zabaras, D., & Konczak, I., (2014). Phenolic composition, antioxidant and enzyme inhibitory activities of *Eryngium bornmuelleri* leaf. *Plant Foods Hum. Nutr.*, *69*(1), 30–36.

42. Shalaby, N. M. M., Abd-Alla, H. I., Aly, H. F., Albalawy, M. A., Shaker, K. H., & Bouajila, J., (2014). Preliminary *in vitro* and *in vivo* evaluation of antidiabetic activity of *Ducrosia anethifolia boiss* and its linear furanocoumarins. *Biomed Res. Int.*, 1–13.
43. Taghizadeh, M., Memarzadeh, M. R., Asemi, Z., & Esmaillzadeh, A., (2015). Effect of the cumin cyminum L. intake on weight loss, metabolic profiles and biomarkers of oxidative stress in overweight subjects: A randomized double-blind placebo-controlled clinical trial. *Ann. Nutr. Metab.*, *66*(2–3), 117–124.
44. Wang, K., Cao, P., Shui, W., Yang, Q., Tang, Z., & Zhang, Y., (2015). Angelica sinensis polysaccharide regulates glucose and lipid metabolism disorder in prediabetic and streptozotocin-induced diabetic mice through the elevation of glycogen levels and reduction of inflammatory factors. *Food Funct.*, *6*(3), 902–909.
45. Ali, M. Y., Jannat, S., Jung, H. A., Jeong, H. O., Chung, H. Y., & Choi, J. S., (2016). Coumarins from *Angelica decursiva* inhibit α-glucosidase activity and protein tyrosine phosphatase 1B. *Chem. Biol. Interact.*, *252*, 93–101.
46. Tovchiga, O. V., (2016). The influence of goutweed (*Aegopodium Podagraria L.*) tincture and metformin on the carbohydrate and lipid metabolism in dexamethasone-treated rats. *BMC Complement. Altern. Med.*, *16*, 235.
47. Bae,U. J., Choi, E. K., Oh, M. R., Jung, S. J., Park, J., Jung, T. S., Park, T. S., Chae, S. W., & Park, B. H., (2016). *Angelica gigas* ameliorates hyperglycemia and hepatic steatosis in C57BL/KsJ- *Db/db* Mice via activation of AMP-activated protein kinase signaling pathway. *Am. J. Chin. Med.*, *44*(8), 1627–1638.
48. Zhang, H., Su, Y., Wang, X., Mi, J., Huo, Y., Wang, Z., Liu, Y., & Gao, Y., (2017). Antidiabetic activity and chemical constituents of the aerial parts of *Heracleum dissectum ledeb. Food Chem.*, *214*, 572–579.
49. Guo, J., Wang, J., Song, S., Liu, Q., Huang, Y., Xu, Y., Wei, Y. X., & Zhang, J., (2016). *Sphallerocarpus gracilis* polysaccharide protects pancreatic β-cells via regulation of the Bax/bcl–2, Caspase–3, Pdx–1 and insulin signaling pathways. *Int. J. Biol. Macromol.*, *93*, 829–836.
50. Aissaoui, A., Zizi, S., Israili, Z. H., & Lyoussi, B., (2011). Hypoglycemic and hypolipidemic effects of *Coriandrum Sativum* L. in meriones shawi rats. *J. Ethnopharmacol.*, *137* (1), 652–661.
51. Grieve, M. *A Modern Herbal.*1 rd ed.; Ed Greenwood: Arcata California, 1995.
52. Peters, C. The proud plant *Petroselinum Crispum*. Technical report of International Open University for Traditional Ayurveda, Unani, Siddha and Tribal Medicine. 2015.
53. Ozsoy-Sacan, O., Yanardag, R., Orak, H., Ozgey, Y., Yarat, A., & Tunali, T., (2006). Effects of Parsley (*Petroselinum Crispum*) extract versus glibornuride on the liver of streptozotocin-induced diabetic rats. *J. Ethnopharmacol.*, *104*(1–2), 175–181.
54. The International Plant Names Index. https://www.ipni.org/. (accessed Jun 22, 2018).
55. Huxley, A. J., Griffiths, M., & Levy, M. The new Royal Horticultural Society dictionary of gardening, 1 rd.; Macmillan Press: Cornell University, 1992.
56. Farahani, H. A. Karimian, D., & Maroufi K., (2012). Effect of vitamin E on seedling growth in parsley *(Petroselinum sativum L.).* Afr. J. Microbiol. Res. *6*(30), 5934–5939.

57. Spraul, M. H., Nitz, S., Drawert, F., Duddeck, H., & Hiegemannt, M., (1992). Crispane and crispanone, two compounds from *petroselinum crispum* with a new carbon @skeleton. *Phytochemistry*, *31*(9), 3109–3111.
58. Ververidis, F., Trantas, E., Douglas, C., Vollmer, G., Kretzschmar, G., & Panopoulos, N., (2007). Biotechnology of flavonoids and other phenylpropanoid-derived natural products. Part I: chemical diversity, impacts on plant biology and human health. *Biotechnol. J.*, *2*(10), 1214–1234.
59. Heim, K. E., Tagliaferro, A. R., & Bobilya, D. J., (2002). Flavonoid antioxidants: Chemistry, metabolism and structure-activity relationships. *J. Nutr. Biochem.*, *13*(10), 572–584.
60. Chaves, D. S., & Frattani, F. S. A. M., (2011). Phenolic chemical composition of *petroselinum crispum* extract and its effect on haemostasis. *Nat. Prod. Commun.*, 961–964.
61. Gadi, D., Bnouham, M., Aziz, M., Ziyyat, A., Legssyer, A., Bruel, A. et al., (2012). Flavonoids purified from parsley inhibit human blood platelet aggregation and adhesion to collagen under flow. *J. Complement. Integr. Med.*, *9*(1).
62. Chaudhary, S. K., Ceska, O., Têtu, C., Warrington, P. J., Ashwood-Smith, M. J., & Poulton, G. A., (1986). Oxypeucedanin, a major furocoumarin in parsley, *Petroselinum Crispum. Planta Med.*, *462*(6), 462–464.
63. Halapas, A., Papalois, A., Stauropoulou, A., Philippou, A., Pissimissis, N., Chatzigeorgiou, A. et al., (2008). *In vivo* models for heart failure research. *In Vivo (Brooklyn)*, *22*(6), 767–780.
64. Yanardag, R., Bolkent, S., Tabakoglu-Oguz, A., & Özsoy-Saçana, Ö., (2003). Effects of *Petroselinum crispum* extract on pancreatic B cells and blood glucose of streptozotocin-induced diabetic rats, *26*(8), 1206–1210.
65. Yanardag, R., Oksel, G., & Ul, G., (2003). Effects of parsley (*Petroselinum Crispum*) on the aorta and heart of stz induced diabetic rats, 1–7.
66. Tunali, T., Yarat, A., Yanardağ, R., Özçelik, F., Özsoy, Ö., Ergenekon, G., & Emekli, N., (1999). Effect of parsley (*Petroselinum crispum*) on the skin of STZ induced diabetic rats. *Phyther. Res.*, *13*(2), 138–141.
67. Eltablawy, N. A., Soliman, H. A., Hamed, M. S., & Division, B., (2015). Antioxidant and antidiabetic role of *Petroselinum crispum* against Stz-induced diabetes in rats., *4*(3), 32–45.
68. Özçelik, F., Yarat, A., Yanardag, R., Tunali, T., Özsoy, Ö., Emekli, N., & Üstüner, A., (2001). Limited effects of parsley (*Petroselinum Crispum*) on protein glycation and glutathione in lenses of streptozotocin-induced diabetic rats. *Pharm. Biol.*, *39*(3), 230–233.
69. Tunev, S. S., Hastey, C. J., Hodzic, E., Feng, S., Barthold, S. W., & Baumgarth, N., (2011). Lymphoadenopathy during lyme borreliosis is caused by spirochete migration-induced specific B cell activation. *PLoS Pathog.*, *7*(5), 20–24.
70. Hozayen, W. G., El-Desouky, M. A., Soliman, H. A., Ahmed, R. R., & Khaliefa, A. K., (2016). Antiosteoporotic effect of *Petroselinum crispum*, *Ocimum basilicum* and *Cichorium intybus L.* in glucocorticoid-induced osteoporosis in rats. *BMC Complement. Altern. Med.*, *16*(1), 165.

71. Akinci, A., Esrefoglu, M., Taslidere, E., & Ates, B., (2017). *Petroselinum crispum* is effective in reducing stress-induced gastric oxidative damage. *Balkan Med. J.*, *34*(1), 53–59.
72. Haidari, F., Keshavarz, S. A., Shahi, M. M., Mahboob, S. A., & Rashidi, M. R., (2011). Effects of parsley (*Petroselinum crispum*) and its flavonol constituents, kaempferol and quercetin, on serum uric acid levels, biomarkers of oxidative stress and liver xanthine oxidoreductase aactivity inoxonate-induced hyperuricemic rats. *Iran. J. Pharm. Res.*, *10*(4), 811–819.
73. Mirzaie, D. N., Moazedi, A. A., & Seyyednejad, S. M., (2010). The role of α-and β - adrenergic receptors in the spasmolytic effects on rat ileum of *Petroselinum crispum Latifolum* (Parsley). *Asian Pac. J. Trop. Med.*, *3*(11), 866–870.
74. Wong, P. Y. Y., & Kitts, D. D., (2006). Food chemistry studies on the dual antioxidant and antibacterial properties of parsley (*Petroselinum crispum*) and cilantro (*Coriandrum sativum*) extracts, *97*, 505–515.
75. Tang, E. L. H., Rajarajeswaran, J., Fung, S., & Kanthimathi, M. S., (2015). *Petroselinum crispum* has antioxidant properties, protects against DNA damage and inhibits proliferation and migration of cancer cells. *J. Sci. Food Agric.*, *95*(13), 2763–2771.
76. Kleinwächter, M., Paulsen, J., Bloem, E., Schnug, E., & Selmar, D., (2015). Moderate drought and signal transducer induced biosynthesis of relevant secondary metabolites in thyme (*Thymus vulgaris*), greater celandine (*Chelidonium majus*) and parsley (*Petroselinum crispum*). *Ind. Crops Prod.*, *64*, 158–166.
77. Seczyk, L., Swieca, M., & Gawlik-Dziki, U., (2015). Changes of antioxidant potential of pasta fortified with parsley (*Petroselinum crispum mill.*) leaves in the light of protein-phenolics interactions. *Acta. Sci. Pol. Technol. Aliment.*, *14*(1), 29–36.
78. Mišan, A., Mimica-Dukić, N., Sakač, M., Mandić, A., Sedej, I., Šimurina, O., & Tumbas, V., (2011). Antioxidant activity of medicinal plant extracts in cookies. *J. Food Sci.*, *76*(9), 1239–1244.
79. Sayed-Ahmad, B., Talou, T., Saad, Z., Hijazi, A., & Merah, O., (2017). The *Apiaceae*: Ethnomedicinal family as source for industrial uses. *Ind. Crops Prod.*, *109*, 661–671.

CHAPTER 9

NUTRACEUTICALS FROM MEXICAN PLANTS AS ANTIBACTERIAL AGENTS

RICARDO GUADALUPE LÓPEZ RAMOS, ANNA ILYINA,
LUIS ENRIQUE COBOS PUC, CRYSTEL ALEYVICK SIERRA RIVERA,
JUAN ALBERTO ASCACIO VALDÉS, and
SONIA YESENIA SILVA BELMARES

Food Research Department, Faculty of Chemistry, Autonomous University of Coahuila, Blvd. Venustiano Carranza, Colony Republic, Zip code 25280, Saltillo, Coahuila, Mexico, E-mail: yesenia_silva@uadec.edu.mx

ABSTRACT

The nutraceutical compounds use in the food industry is diverse; some are part of functional foods while others prolong foods shelf life. In Mexico, many medicinal plants are used to flavor and prolong the shelf life of foods, some of which are used to treat infectious diseases caused by contaminated foods, making them of great importance to the society. One of the main uses of the plants used in the country is related to treatment and prevention of infectious diseases, therefore the most could be used for the development of functional foods since have nutraceutical properties. This review is focused in the antibacterial component presents in those plants used in Mexico for their medicinal properties, and for food elaboration. Coumarins, terpenes, flavonoids, and alkaloids have been identified as compounds with antibacterial effect in medicinal Mexican plants. On the other hand, extracts of Mexican plants exhibit greater efficacy on gram-negative than gram-positive bacteria. For this reason, they could be used to develop new nutraceuticals by its addition during formulation foods.

9.1 INTRODUCTION

The major function of the foods is providing the nutrients necessary to maintain the health of the human body. Diets abundant in fruits and vegetables reduce the risk of infectious diseases and prevent diseases related to aging [1–5]. Nowadays, the selection of food is focused on nutritional as well as functional aspect. These foods are classified into functional, medical, supplements, and nutraceuticals [6]. In 1989, Stephen DeFelice defined the nutraceutical term as a food or compound that has benefits for the human organism, including to phytochemicals with biological activity [3, 7]. Currently, misuse of antibiotics it has caused the apparition of bacterial strains resistant to classical antibiotics, so there is a growing problem in the control of infections [8, 9]. In Mexico, clinical infectious occurs by ingestion of contaminated food with pathogenic microorganisms [10]. For this reason, some plants with antimicrobial activity are added to meals in order to conserve its useful life and improve the flavor. Additionally, the Mexican population uses medical plants to treat infectious diseases. Mexico possesses an extraordinary flora that includes 22,000–31,000 species and 17% are medical plants [11]. For these reasons, in Mexico and the world, there is an increase in the research focused in found new antimicrobial compounds from the plant [12, 13]. These investigations have a social, economic, and cultural impact based on that strengthen the development of indigenous communities through the cultivation, distribution, and sale of medicinal plants [14, 15]. The antimicrobial effect of plants has a great potential for foods nutraceutical development to prevent infections by contaminated meals.

9.2 MEXICAN PLANTS AS ANTIBACTERIAL AGENTS

In Mexico, there exists a great variety of plants that have a different use, many of them share the antimicrobial effect, so in the next section, we described those that are used for this purpose.

9.2.1 MENTHA SPICATA

Mentha spicata is known as peppermint or mint and is one of the most used medicinal plants in the world. Among the uses of this plant is the production of beverages such as tea, the production of perfumes, confectionery, and pharmaceuticals. The limonene is a major compound from the essential oil of this plant and shows an effect against gram-positive bacteria [16]; also, this plant synthesizes the carvone [17]. Additionally, limonene can interact with other compounds through a synergism to potentiate the antimicrobial effect [18].

In a study carried out with *M. spicata*, it has been observed that the essential oil of cultures in combination with potassium increases the inhibition of gram-positive and gram-negative bacteria [17].

9.2.2 LARREA TRIDENTATA

This plant is a shrub that in Mexico is known as "gobernadora" and belongs to the family Zygophyllaceae, is endemic from the northern of Mexico and is used for its medicinal properties. This plant is used in alternative treatments of degenerative, menstrual, and infectious diseases. Leaves and stems of this plant contain phenolic compounds such as flavonoids, as well as lignans and triterpenes [19, 20].

Recently, was reported that the extracts (0–300 μg) of *L. tridentata* reduces the bacterial replication of *S. aureus* [21]. In another work, three lignans and four flavonoids of *L. tridentata* were isolated and tested against 16 strains, some of them multidrug - resistant. The tests revealed that *Staphylococcus aureus* and *Mycobacterium tuberculosis, Enterobacter cloacae* and *M. tuberculosis*, *S. aureus*, *Enterococcus faecalis*, *Escherichia coli*, *E. cloacae*, and *M. tuberculosis*, and *S. aureus* and *E. faecalis* were susceptible to dihydroguaiarétic acid, 4-epi-larreatricina, 3'-Demethoxy–6-O-demethylisoguaiacin, and 5,4'-dihydroxy–3,7,8,3'-tetramethoxyflavone, respectively [22].

The effect of Nordihydroguaiaretic acid on strains of *S. aureus* was confirmed in methicillin-sensitive *S. aureus* (MSSA) and methicillin-resistant *S. aureus* (MRSA) strains. Additionally, this compound shows an antioxidant effect that destabilizes the microbial membranes so it could use for the development of nutraceuticals [23].

9.2.3 ACHILLEA MILLEFOLIUM

Achillea millefolium is an aromatic plant known as "milenrrama" and belongs to the family Astaraceae. This plant is used in traditional Mexican medicine because has inflammatory, antimicrobial, and antiviral properties. Some compounds identified in this plant are flavonoids, coumarins, alkaloids, lactones, and tannins [24].

In recent studies, essential oils from leaves of *A. millefolium* inhibited the growth of *Bacillus cereus*, *Enterococcus faecalis*, *S. aureus*, *Escherichia coli*, *Pseudonomas aeruginosa*, *Proteus mirabilis*, *Salmonella typhimurium* and *Citrobacter freundii* with higher potency than the reference antibiotics [25]. Also, has been observed that the essential oil has antifungal effects [26]. Additionally, in the tincture of *A. millefolium* were detected two antibacterial fractions by the bioautography method. These fractions reduced the growth of *S. aureus* (ATCC 29213), *S. aureus* (methicillin-resistant), *Streptococcus epidermidis* and *E. coli* [27].

In other studies, the antibacterial effect of the hydro-alcoholic extracts of *A. millefolium* on *Streptococcus mutans*, *Lactobacillus rhamnosus*, and *Actinomyces viscosus* was observed [28].

9.2.4 LIPPIA GRAVEOLENS

Lippia graveolens is known as "Mexican oregano," an aromatic plant which is used to provide flavor and aroma to food. The essential oils of this plant are used in the pharmaceutical industry by its nutraceutical properties, due to its antioxidant and antibacterial effects [29–31]. This plant is used in traditional Mexican medicine to treat gastrointestinal and respiratory infections.

The *L. graveolens* essential oils contain carvacrol, α-terpinyl acetate, thymol as well as β-pinene, and these compounds has been associated with antibacterial activity against *Vibrio cholerae*, *Salmonella tyhpi*, and *Yersinia enterocolitica* [32, 33].

Also, thymol and carvacrol have antibacterial effects against *Escherichia coli*, *Staphylococcus aureus*, *Staphylococcus epidermidis*, *Streptococcus faecalis* and *Proteus vilgaris*, with MIC values of 0.25–0.83 μl/ml [34]. Techniques to protect nutraceutical compounds should be used to prevent the loss of biological activity. The microencapsulation technique is an example since it allows control the release of nutraceutical compounds and it has been demonstrated that some materials to encapsulate several compounds as p-cymene, thymol, and carvacrol have bactericidal effects against pathogenic strains [30]. Additionally, thymol and carvacrol can produce a synergistic effect with nutraceutical substances as well as great bacterial inhibition [35].

9.2.5 *ALIUM SATIVUM*

Alium sativum is cultivated in Mexico and is known as garlic, this crop is one of the most profitable, so it grows about 5,451 ha. The states with the highest production are Zacatecas, Guanajuato, Sonora, Puebla, Baja California and Aguascalientes [36]. Extracts of *A. sativum* present an effect on *Vibrio cholerae* and act synergistically with antibiotics [37]. Also, nanoparticles with garlic extracts exhibit a high antimicrobial effect, especially against human pathogenic microorganisms such as S*treptococcus faecalis* (ATCC: 29212), *Bacillus cereus* (ATCC: 10702), *Escherichia coli* (ATCC: 25922), and *Shigella flexneri* (KZN) [38]. The garlic juice has high activity against *Clostridium difficile in vitro* [39]. The mixture of garlic extracts with higher amounts of diallyl trisulfide potentiate their antimycobacterial activity [40].

9.2.6 *ALIUM CEPA*

Alium cepa is known as onion and is used in Mexico to improve the taste of meals. The essential oils of this plant are rich in sulfur compounds such as dipropyl disulfide and dipropyl trisulfide which display antibacterial activity against *Escherichia coli* O157: H7 (MIC = 5.13 g L^{-1}), *Salmonella choleraesuis* (MIC = 1.28 g L^{-1}), *Listeria monocytogenes* (MIC = 2.56 g L^{-1}) and *Staphylococcus aureus* (MIC = 5.26 g L^{-1}) [41]. On the other hand, a potent antimicrobial activity of methanolic and aqueous extracts of *Allium cepa* in *Bacillus subtilis* and *Pseudomonas aeruginosa* was observed using the disc diffusion method and the minimum inhibitory concentration by the microplate method [42].

9.2.7 TAGETES LUCIDA

Tagetes lucida belongs to the Asteraceae family and is known as Mexican tarragon. This is an important plant for its nutritional and medicinal properties. This plant has an antibacterial effect on gram-positive and negative bacteria since the MeOH/CH_2Cl_2 extract inhibits the growth of *E. coli*, *P. mirabilis*, *K. pneumoniae*, *Salmonella sp.* and *Shigella sp.* [43]. Additionally *T. lucida* it has been used as an antibacterial agent in respiratory tract infections [44]. Extracts from this plant, also, exhibits antibacterial actions on *Escherichia coli*, *Salmonella enteritidis*, *Salmonella typhi*, *Shigella dysenteriae* and *Shigella flexneri* [45].

9.2.8 PERSEA AMERICANA

Persea americana Mill. (Lauraceae), is native to Mexican tropical areas, currently is cultivated in other regions of Latin America, the United States, and Europe [46]. The crude extracts of stem bark and the butanolic fraction from *Persea americana* show effect on *Bacillus cereus* strains involved in food poisoning at 25 mg/ml and 10 mg/ml respectively [47]. Additionally, the avocado seed extract enriched whit acetogenin has an antilisterial effect (*Listeria monocytogenes*) similar to commercial antimicrobial Avosafe® [48]. *P. americana* chloroformic extract inhibits the growth of *M. tuberculosis* H37Rv, M. tuberculosis MDR SIN 4 isolate, three *M. tuberculosis* H37Rv mono-resistant reference strains and four non tuberculosis mycobacteria (*M. fortuitum*, *M. avium*, *M. smegmatis* and *M. absessus*) showing MIC values ≤ 50 μg/ml [49]. Ethanolic extracts show a great inhibitory effect against *Streptococcus mutans* and *Porphyromonas gingivalis* [50].

9.2.9 ORIGANUM MAJORANA

Origanum majorana is a plant used in Mexico to flavor foods, some studies have evaluated its antibacterial effect. Interestingly, essential oils have higher effect on gram-negative bacteria than gram-positive bacteria. Compared with synthetic antibiotics, essential oils were more effective against *E. coli*, *L. innocua* and *S. enteridis* [51]. Additionally, this plant has an antioxidant effect [52]. The essential oil of this plant is not mutagenic

[53]. Another study demonstrated that the essential oil of *Origanum majorana* displays antimicrobial activity against 25 bacterial and fungal strains [54].

9.2.10 ROSMARINUS OFFICINALIS

Rosmarinus officinalis L. is a plant used in traditional Mexican medicine and food processing [55]. Extract display a great efficacy against *Escherichia coli* O157:H7, *Salmonella enteritidis*, *Salmonella typhi*, *Yersinia enterocolitica* and *Listeria monocytogenes* [56]. Some essential oil blends of *Cinnamomum zeylanicum*, *Daucus carota*, *Eucalyptus globulus*, and *Rosmarinus officinalis* show an effect against fourteen gram-positive and gram-negative strains, including some antibiotic-resistant [57]. *Rosmarinus officinalis* L. extract has an inhibitory effect of *Candida albicans*, *Staphylococcus aureus*, *Enterococcus faecalis*, *Streptococcus mutans* and *Pseudomonas aeruginosa* on monomicrobial biofilms responsible for infections in the oral cavity as well as in other regions of the body [55].

9.2.11 SALVIA OFFICINALIS

S. officinalis belongs to Lamiaceae family, in Mexico oral infusion is used to treat respiratory and gastrointestinal tract infections [58]. *Salvia officinalis* extract reduces the growth of *Enterococcus faecalis* [59]. Also, reduces the dental-bacterial plaque since bacterial colonies number reduced of 3900 to 300 [60]. On the other hand, the ethanolic extract of leaves of *Salvia officinalis* shows antibacterial activity against *Bacillus cereus*. The fractionation of this extract led to the isolation of the diterpene, methyl carnosate [61].

9.3 COMPOUNDS RELATED TO ANTIBACTERIAL EFFECT

In Mexico, there are a large number of plants containing compounds with antibacterial effect, some act synergistically others act individually. Some compounds identified in different investigations described below (Table 9.1). These are related to antibacterial effect against different strains.

TABLE 9.1 Secondary Metabolites From Plants with Antibacterial Activity

Plant	Identified compounds	Bacteria	Reference
Tagetes lucida	Seven coumarins: 7,8-dihydroxycoumarin, umbelliferone (7-hydroxycoumarin), scoparone (6,7-dimethoxycoumarin), esculetin (6,7-dihydroxycoumarin), 6-hydroxy–7-methoxycoumarin, herniarin (7-methoxycoumarin), and scopoletin (6-methoxy–7-hydroxycoumarin). Three flavonoids: patuletin, quercetin, and quercetagetin	*Bacillus subtilis, Escherichia coli, Proteus mirabilis, Klebsiella pneumoniae, Salmonella typhi, Salmonella sp., Shigella boydii, Shigella sp., Enterobacter aerogenes, Enterobacter agglomerans, Sarcina lutea, Staphylococcus epidermidis, Staphylococcus aureus, Yersinia enterolitica, Vibrio cholera*	[43]
Persea americana var. drymifolia	Defensin	*Escherichia coli, Staphylococcus aureus*, and *Candida albicans*	[62]
Juglans regia L.	α-pinene, β-pinene, β-caryophyllene, germacrene D and limonene.	*Staphylococcus epidermidis* MTCC–435, Bacillus subtilis MTCC–441, *Staphylococcus aureus, Proteus vulgaris* MTCC–321, *Pseudomonas aeruginosa* MTCC–1688, *Salmonella typhi, Shigella dyssenteriae, Klebsiella pneumoniae* and *Escherichia coli*	[63]
Mentha piperita, Salvia officinalis, Rosmarinus officinalis.	Menthol, thymol, and carvacrol.	*Streptococcus mutans*	[64]

9.4 MECHANISM OF ANTIBACTERIAL ACTIVITY OF NUTRACEUTICALS

Nutraceuticals are contained in plant extracts and the antibacterial effect may vary depending on each compound. Table 9.2 summarizes some families of compounds having antibacterial activity and their mechanism of action.

TABLE 9.2 Compounds Identified in Mexican Plants and Their Mechanism of Action Antibacterial

Compound	Action Mechanism	References
Flavonoids	Inhibits nucleic acids synthesis	[65, 66]
	Bacterial cell membrane damage	
	Inhibits cytoplasmic membrane function	
	Inhibits energy metabolism	
Terpenes	Inhibits the protective enzymes of bacterial membrane	[67, 68]
	Bacterial cell membrane damage	
Alkaloids	Inhibits nucleic acids synthesis Bacterial cell membrane damage	[69]

9.5 FINAL COMMENTS

Despite the availability of synthetic compounds, the search of natural compounds to lower cost and few side effects are desirable for human beings. Therefore, there is a growing interest in preventive medicine promoting the use of nutraceuticals by their important nutritional values and effects on the health. The purpose of this review is to emphasize the main components and their mechanisms of action of some medicinal plants such as *Mentha spicata, Larrea tridentata, Achillea millefolium, Lappia graveolens, Alium sativum, Alium cepa, Tagetes lucida, Persea americana, Origanum majorana, Rosmarinus officinalis, Salvia officinalis*, among others since the compounds containing these plants are promising for the development of nutraceuticals. These compounds found in medicinal plants should be evaluated in vivo assays to guarantee their safety and subsequently carry out the successful implementation to the market.

KEYWORDS

- medicinal plants
- nutraceuticals
- synthetic compounds

REFERENCES

1. Costa, A. G. V., Garcia-Diaz, D. F., Jimenez, P., & Silva, P. I., (2013). Bioactive compounds and health benefits of exotic tropical red–black berries. *J. Funct. Foods*, *5*(2), 539–549.
2. Ostan, R., Béné, M. C., Spazzafumo, L., Pinto, A., Donini, L. M., Pryen, F. et al., (2016). Impact of diet and nutraceutical supplementation on inflammation in elderly people. Results from the RISTOMED study, an open-label randomized control trial. *Clin. Nutr.*, *35*(4), 812–818.
3. Santini, A., Tenore, G. C., & Novellino, E., (2017). Nutraceuticals: A paradigm of proactive medicine. *Eur. J. Pharm. Sci.*, *96*, 53–61.
4. Biruete, G. A., Juárez, H. E., Sieiro, O. E., Romero, V. R., & Silencio, B. J., (2009). The nutraceuticals. What is convenient to know. *Rev. Mex. Pediatría*, *76*(3), 136–145.
5. Cruzado, M., & Cedrón, J. C., (2012). Nutracéuticos, alimentos funcionales y su producción. *Rev. Química PUCP*, *26*(1–2), 33–36.
6. Brown, A. W. *Encyclopedia of Meat Sciences:* Human nutrition/nutraceuticals, 1st ed. Academic Press, 2014.
7. De Felice, S. L., (1995). The nutraceutical revolution: Its impact on food industry R&D. *Trends Food Sci. Technol.*, *6*(2), 59–61.
8. Pandey, S., (2015). Preliminary phytochemical screening and in vitro antibacterial activity of bauhinia variegata linn. against human pathogens. *Asian Pacific J. Trop. Dis.*, *5*(2), 123–129.
9. Islam, R., Rahman, M. S., Hossain, R., Nahar, N., Hossin, B., Ahad, A., & Rahman, S. M., (2015). Antibacterial activity of combined medicinal plants extract against multiple drug resistant strains. *Asian Pacific J. Trop. Dis.*, *5*, 151–154.
10. Verraes, C., Van Boxstael, S., Van Meervenne, E., Els, V. C., Butaye, P., Catry, B., De Schaetzen, M. A. et al., (2013). Antimicrobial resistance in the food chain: A review. *Int. J. Environ. Res. Public Health*, *10*(7), 2643–2669.
11. Conabio. *Estrategia Mexicana Para La Conservación Vegetal 2012–2030*; México, 2012.
12. Juárez-Vázquez, M., Del, C., Carranza-Álvarez, C., Alonso-Castro, A. J., González-Alcaraz, V. F., Bravo-Acevedo, E. et al., (2013). Ethnobotany of medicinal plants used in xalpatlahuac, Guerrero, México. *J. Ethnopharmacol.*, *148*(2), 521–527.
13. Hossain, M. A., Al-Musalami, A. H. S., Akhtar, M. S., & Said, S., (2014). A comparison of the antimicrobial effectiveness of different polarities crude extracts from the

leaves of adenium obesum used in omani traditional medicine for the treatment of microbial infections. *Asian Pacific J. Trop. Dis.*, *4*(2), 934–937.
14. Sola, B., (2016). Escasa investigación científica en plantas medicinales http://www.conacytprensa.mx/index.php/noticias/reportaje/5404-escasa-investigacion-cientifica-en-plantas-medicinales.
15. Salih, E. Y. A., Kanninen, M., Sipi, M., Luukkanen, O., Hiltunen, R., & Vuorela, H., (2017). Tannins, flavonoids and stilbenes in extracts of African savanna woodland trees terminalia brownii, terminalia laxiflora and anogeissus leiocarpus showing promising antibacterial potential. *South African J. Bot.*, *108*, 370–386.
16. Telci, I., Demirtas, I., Bayram, E., Arabaci, O., & Kacar, O., (2010). Environmental variation on aroma components of pulegone/piperitone rich spearmint (*Mentha Spicata* L.). *Ind. Crops Prod.*, *32*(3), 588–592.
17. Chrysargyris, A., Xylia, P., Botsaris, G., & Tzortzakis, N., (2017). Antioxidant and antibacterial activities, mineral and essential oil composition of spearmint (*Mentha Spicata* L.) affected by the potassium levels. *Ind. Crops Prod.*, *103*, 202–212.
18. Hernández-Ochoa, L., Gonzales-Gonzales, A., Gutiérrez-Mendez, N., Muñoz-Castellanos, L. N., & Quintero-Ramos, A., (2011). Study of the antibacterial activity of chitosan–based films prepared with different molecular weights including spices essential oils and functional extracts as antimicrobial agents. *Rev. Mex. Ing. Química*, *10*(3), 455–463.
19. Jitsuno, M., & Mimaki, Y., (2010). Triterpene glycosides from the aerial parts of larrea tridentata. *Phytochemistry*, *71*(17), 2157–2167.
20. Martins, S., Aguilar, C. N., Teixeira, J. A., & Mussatto, S. I., (2012). Bioactive compounds (Phytoestrogens) recovery from larrea tridentata leaves by solvents extraction. *Sep. Purif. Technol.*, *88*, 163–167.
21. Snowden, R., Harrington, H., Morrill, K., Jeane, L., Garrity, J., Orian, M. et al., (2014). A comparison of the anti-staphylococcus aureus activity of extracts from commonly used medicinal plants. *J. Altern. Complement. Med.*, *20*(5), 375–382.
22. Favela-Hernández, J., García, A., Garza-González, E., Rivas-Galindo, V., & Camacho-Corona, M., (2012). Antibacterial and antimycobacterial lignans and flavonoids from larrea tridentata. *Phyther. Res.*, *26*(12), 1957–1960.
23. Cunningham-Oakes, E., Soren, O., Moussa, C., Rathor, G., Liu, Y., & Coates, A., (2015). Nordihydroguaiaretic acid enhances the activities of aminoglycosides against methicillin- sensitive and resistant *Staphylococcus aureus in vitro* and *in vivo* front. *Microbiol.*, *6*.
24. Krapp, K., & Longe, J. L., (2006). *Enciclopedia de Las Medicinas Alternativas*; OCEANO: Barcelona.
25. Kazemi, M., (2015). Chemical composition and antimicrobial, antioxidant activities and anti-inflammatory potential of *Achillea millefolium* L., *Anethum graveolens* L and *Carum copticum* L. essential oils. *J. Herb. Med.*, *5*(4), 217–222.
26. El-Kalamouni, C., Venskutonis, P., Zebib, B., Merah, O., Raynaud, C., & Talou, T., (2017). Antioxidant and antimicrobial activities of the essential oil of achillea millefolium L. grown in France. *Med.*, *4*(2).
27. Jesionek, W., Móricz, Á. M., Ott, P. G., Kocsis, B., Horváth, G., & Choma, I. M., (2015). TLC-direct bioautography and LC/MS as complementary methods in identi-

fication of antibacterial agents in plant tinctures from the asteraceae family. *J. AOAC Int.*, *98*(4), 857–861.

28. Kermanshah, H., Kamangar, S., Arami, S., Kamalinegad, M., Karimi, M., Mirsalehian, A., Jabalameli, F., & Fard, M., (2014). The effect of hydro alcoholic extract of seven plants on cariogenic bacteria--an *in vitro* evaluation. *J. Oral Heal. Dent. Manag.*, *13*(2), 395–401.
29. García-Pérez, E., Castro-Álvarez, F. F., Gutiérrez-Uribe, J. A., & García-Lara, S., (2012). Revision of the production, phytochemical composition, and nutraceutical properties of Mexican oregano. *Rev. Mex. Ciencias. Agrícolas*, *3*(2), 339–353.
30. Arana-Sánchez, A., Estarrón-Espinosa, M., Obledo-Vazquéz, E. N., Padilla-Camberos, E., Silva-Vazquéz, R., & Lugo-Cervantes, E., (2010). Antimicrobial and antioxidant activities of Mexican oregano essential oils (*Lippia graveolens* H. B. K.) with different composition when microencapsulated Inβ-Cyclodextrin. *Lett. Appl. Microbiol.*, *50*(6), 585–590.
31. Pino, J., Rosado, A., Baluja, R., & Borges, P., (1989). Analysis of the essential oil of Mexican oregano (*Lippia graveolens* HBK). *Mol. Nutr. Food Res.*, *33*(3), 289–295.
32. Hernández, T., Canales, M., Avila, J. G., García, A. M., Meraz, S., Caballero, J. et al., (2009). Composition and antibacterial activity of essential oil of *Lippia graveolens* H.B.K. (Verbenaceae). *Boletín Latinoam. y del Caribe Plantas Med. y Aromáticas*, *8*(4), 295–300.
33. Bueno-Durán, A. Y., Cervantes-Martínez, J., & Obledo-Vázquez, E. N., (2014). Composition of essential oil from *Lippia graveolens*. Relationship between spectral light quality and thymol and carvacrol content. *J. Essent. Oil Res.*, *26*(3), 153–160.
34. Salgueiro, L. R., Cavaleiro, C., Gonçalves, M. J., & Proença da Cunha, A., (2003). Antimicrobial activity and chemical composition of the essential oil of *Lippia graveolens* from guatemala. *Planta. Med.*, *69*(1), 80–83.
35. Hernández-Hernández, E., Regalado-González, C., Vázquez-Landaverde, P., Guerrero-Legarreta, I., & García-Almendárez, B., (2014). Microencapsulation, chemical characterization, and antimicrobial activity of Mexican (*Lippia graveolens* H.B.K.) and European (*Origanum vulgare* L.) oregano essential oils. *Sci. World J.*, 1–12.
36. Delgado-Ortiz, J., Ochoa-Fuentes, Y., Cerna-Chávez, E., Beltrán-Beache, M., Rodríguez-Guerra, R., Aguirre-Uribe, L. et al., (2016). *Fusarium* species associated with basal rot of garlic in North Central Mexico and its pathogenicity. *Rev. Argent. Microbiol.*, *48*(3), 222–228.
37. Bruns, M., Kakarla, P., Floyd, J., Mukherjee, M., Ponce, R., Garcia, J. et al., (2017). Modulation of the multidrug efflux pump EmrD–3 from *Vibrio cholerae* by *Allium sativum* extract and the bioactive agent allyl sulfide plus synergistic enhancement of antimicrobial susceptibility by *A. sativum* extract. *Arch. Microbiol.*, *199*(8), 1103–1112.
38. Otunola, G., Afolayan, A., Ajayi, E., & Odeyemi, S., (2017). Characterization, antibacterial and antioxidant properties of silver nanoparticles synthesized from aqueous extracts of *Allium sativum*, *Zingiber officinale*, and *Capsicum frutescens*. *Pharmacogn. Mag.*, *13*(50), 201–208.
39. Roshan, N., Riley, T., & Hammer, K., (2017). Antimicrobial activity of natural products against *Clostridium difficile in Vitro*. *J. Appl. Microbiol.*, *123*(1), 92–103.

40. Oosthuizen, C., Arbach, M., Meyer, D., Hamilton, C., & Lall, N., (2017). Diallyl polysulfides from *Allium sativum* as immunomodulators, hepatoprotectors, and antimycobacterial agents. *J. Med. Food, 20*(7), 685–690.
41. Karasinski, J., Wrobel, K., Corrales Escobosa, A., Konopka, A., Bulska, E., & Wrobel, K., (2017). *Allium cepa* L. response to sodium selenite (Se(IV)) studied in plant roots by a LC-MS-based proteomic approach. *J. Agric. Food Chem., 65*(19), 3995–4004.
42. Kaur, G., Gupta, V., Christopher, A., Bansal, R., & Bansal, P., (2017). Kitchen phytochemicals from *Allium cepa* - their role in multidrug resistance. *Pak. J. Pharm. Sci., 30*(3), 789–792.
43. Céspedes, C., Avila, J., Martínez, A., Serrato, B., Calderón-Mugica, J., & Salgado-Garciglia, R., (2006). Antifungal and antibacterial activities of Mexican tarragon (*Tagetes lucida*). *J. Agric. Food Chem., 54*(10), 3521–3527.
44. Caceres, A., Alvarez, A., Ovando, A., & Samayoa, B., (1991). Plants used in Guatemala for the treatment of respiratory diseases. 1. Screening of 68 plants against gram-positive bacteria. *J. Ethnopharmacol., 31*(2), 193–208.
45. Caceres, A., Cano, O., Samayoa, B., & Aguilar, L., (1990). Plants used in Guatemala for the treatment of gastrointestinal disorders. 1. Screening of 84 plants against Enterobacteria. *J. Ethnopharmacol, 30*(1), 55–73.
46. Ramos-Jerz, M., Del, R., Villanueva, S., Jerz, G., Winterhalter, P., & Deters, A. M., (2013). *Persea Americana* mill. Seed: Fractionation, characterization, and effects on human keratinocytes and fibroblasts. *Evidence-Based Complement. Altern. Med.*, pp. 12.
47. Akinpelu, D., Aiyegoro, O., Akinpelu, O., & Okoh, A., (2014). Stem bark extract and fraction of persea americana (Mill.) exhibits bactericidal activities against strains of *Bacillus cereus* associated with food poisoning. *Molecules, 20*(1), 416–429.
48. Salinas-Salazar, C., Hernández-Brenes, C., Rodríguez-Sánchez, D., Castillo, E., Navarro-Silva, J., & Pacheco, A., (2016). Inhibitory activity of avocado seed fatty acid derivatives (Acetogenins) against *Listeria monocytogenes*. *J. Food Sci., 82*(1), 134–144.
49. Jiménez-Arellanes, A., Luna-Herrera, J., Ruiz-Nicolás, R., Cornejo-Garrido, J., Tapia, A., & Yépez-Mulia, L., (2013). Antiprotozoal and antimycobacterial activities of *Persea Americana* Seeds. *BMC Complement. Altern. Med., 13* (109), 1-5.
50. Rosas-Piñón, Y., Mejía, A., Díaz-Ruiz, G., Aguilar, M., Sánchez-Nieto, S., & Rivero-Cruz, J., (2012). Ethnobotanical survey and antibacterial activity of plants used in the altiplane region of Mexico for the treatment of oral cavity infections. *J. Ethnopharmacol., 141*(3), 860–865.
51. Olfa, B., Mariem, A., Salah, A., & Mouhiba, B., (2016). Chemical content, antibacterial and antioxidant properties of essential oil extract from Tunisian *Origanum Majorana* L. cultivated under saline condition. *Pak. J. Pharm. Sci., 29*(6), 1951–1958.
52. García-Risco, M., Mouhid, L., Salas-Pérez, L., López-Padilla, A., Santoyo, S., Jaime, L. et al., (2017). Biological activities of *Asteraceae* (*Achillea Millefolium* and *Calendula officinalis*) and Lamiaceae (*Melissa officinalis* and *Origanum majorana*) plant extracts. *Plant Foods Hum. Nutr., 72*(1), 96–102.
53. Dos Santos, D. A., Klein-Júnior, L. C., Machado, M. S., Guecheva, T. N., Dos Santos, L. D., Zanette, R. A. et al., (2016). *Origanum Majorana* essential oil lacks mutagenic activity in the salmonella/microsome and micronucleus assays. *Sci. World J.*, p. 7.

54. Hajlaoui, H., Mighri, H., Aouni, M., Gharsallah, N., & Kadri, A., (2016). Chemical composition and in vitro evaluation of antioxidant, antimicrobial, cytotoxicity and anti-acetylcholinesterase properties of tunisian *Origanum majorana* L. essential oil. *Microb. Pathog.*, *95*, 86–94.
55. De Oliveira, J., De Jesus, D., Figueira, L., De Oliveira, F., Pacheco, S. C., Camargo, S. et al., (2017). Biological activities of *Rosmarinus officinalis* L. (Rosemary) extract as analyzed in microorganisms and cells. *Exp. Biol. Med.*, *242*(6), 625–634.
56. Santomauro, F., Sacco, C., Donato, R., Bellumori, M., Innocenti, M., & Mulinacci, N., (2017). The antimicrobial effects of three phenolic extracts from *Rosmarinus officinalis* L., *Vitis vinifera* L. and *Polygonum cuspidatum* L. on food pathogens. *Nat. Prod. Res.*, 1–7.
57. Brochot, A., Guilbot, A., Haddioui, L., & Roques, C., (2017). Antibacterial, antifungal, and antiviral effects of three essential oil blends. *Open Microbiol.*, *6*(4), 1-6.
58. Vogl, S., Picker, P., Mihaly-Bison, J., Fakhrudin, N., Atanasov, A., Heiss, E. et al., (2013). Ethnopharmacological *in vitro* studies on Austria's folk medicine--an unexplored lore *in vitro* anti-inflammatory activities of 71 Austrian traditional herbal drugs. *J. Ethnopharmacol.*, *149*(3), 750–771.
59. Guneser, M., Akbulut, M., & Eldeniz, A., (2016). Antibacterial effect of chlorhexidine-cetrimide combination, *Salvia officinalis* plant extract and octenidine in comparison with conventional endodontic irrigants. *Dent. Mater. J.*, *35*(5), 736–741.
60. Beheshti-Rouy, M., Azarsina, M., Rezaie-Soufi, L., Alikhani, M. Y., Roshanaie, G., & Komaki, S., (2015). The antibacterial effect of sage extract (*Salvia officinalis*) mouthwash against *streptococcus* mutans in dental plaque: A randomized clinical trial. *Iran. J. Microbiol.*, *7*(3), 173–177.
61. Climati, E., Mastrogiovanni, F., Valeri, M., Salvini, L., Bonechi, C., Mamadalieva, N. et al., (2013). Methyl carnosate, an antibacterial diterpene isolated from *Salvia officinalis* leaves. *Nat. Prod. Commun.*, *8*(4), 429–430.
62. Guzmán-Rodríguez, J. J., López-Gómez, R., Suárez-Rodríguez, L. M., Salgado-Garciglia, R., Rodríguez-Zapata, L. C., Ochoa-Zarzosa, A. et al., (2013). Antibacterial activity of defensin PaDef from avocado fruit (*Persea Americana* Var. drymifolia) expressed in endothelial cells against *Escherichia coli* and *Staphylococcus aureus*. *Biomed Res. Int. 2013*, 1-9.
63. Rather, M., Dar, B., Dar, M., Wani, B., Shah, W., Bhat, B. et al., (2012). Chemical composition, antioxidant and antibacterial activities of the leaf essential oil of *Juglans regia* L. and its constituents. *Phytomedicine*, *19*(13), 1185–1190.
64. Tardugno, R., Pellati, F., Iseppi, R., Bondi, M., Bruzzesi, G., & Benvenuti, S., (2017). Phytochemical composition and *in vitro* screening of the antimicrobial activity of essential oils on oral pathogenic bacteria. *Nat. Prod. Res.*, 1–8.
65. Cushnie, T. P. T., & Lamb, A. J., (2005). Antimicrobial activity of flavonoids. *Int. J. Antimicrob. Agents*, *26*(5), 343–356.
66. He, M., Wu, T., Pan, S., & Xu, X., (2014). Antimicrobial mechanism of flavonoids against *Escherichia coli* ATCC 25922 by model membrane study. *Appl. Surf. Sci.*, *305*, 515–521.

67. Trombetta, D., Castelli, F., Sarpietro, M. G., Venuti, V., Cristani, M., Daniele, C. et al., (2005). Mechanisms of antibacterial action of three monoterpenes. *Antimicrob. Agents Chemother.*, *49*(6), 2474–2478.
68. Zengin, H., & Baysal, A. H., (2014). Antibacterial and antioxidant activity of essential oil terpenes against pathogenic and spoilage-forming bacteria and cell structure-activity relationships evaluated by SEM microscopy. *Molecules*, *19*(11), 17773–17798.
69. Cushnie, T. P. T., Cushnie, B., & Lamb, A. J., (2014). Alkaloids: An overview of their antibacterial, antibiotic-enhancing and antivirulence activities. *Int. J. Antimicrob. Agents.*, *44*(5), 377–386.

CHAPTER 10

RELEVANT MARINE BIOMASS AS FEEDSTOCK FOR APPLICATION IN THE FOOD INDUSTRY: AN OVERVIEW

DULCE G. ARGÜELLO-ESPARZA,[1] HÉCTOR A. RUIZ,[1,2] CRISTÓBAL N. AGUILAR,[1] DIANA JASSO DE RODRÍGUEZ,[3] BARTOLOMEU W. S. SOUZA,[4] and ROSA M. RODRÍGUEZ-JASSO[1,2]

[1] *Biorefinery Group, Food Research Department, Faculty of Chemical Sciences, Autonomous University of Coahuila, 25280, Saltillo, Coahuila, Mexico, Tel.: (+52) 844 416 12 38, E-mail: rrodriguezjasso@uadec.edu.mx*

[2] *Cluster of Bioalcohols, Mexican Centre for Innovation in Bioenergy (Cemie-Bio), Mexico*

[3] *Departamento de Fitomejoramiento, Universidad Autonoma Agraria Antonio Narro, Calzada Antonio Narro No. 1923, Colonia Buenavista, 25315, Saltillo, Coahuila, Mexico*

[4] *Departamento de Engenharia de Pesca, Universidade Federal do Ceará/UFC, 60455-970, Fortaleza-Ceará, Brazil*

ABSTRACT

Seaweeds, fish waste (tail, head, fins, fins, and scales), shrimp residues (head, tail, and peel), and other waste of marine species such as jellyfish and crustacean residues are marine biomasses that can be used for human consumption by exploiting its nutritional components. Seaweeds contain proteins, minerals, vitamins, lipids, and polysaccharides; shrimp wastes contribute with chitin and chitosan, while fish waste contains proteins, omega fats, and collagen. In addition to their extensive use in the food industry, a huge variety of these biomasses have in common the ability to act as thickening and gelling improving the rheological characteristics of certain products. These marine biomasses also have relevant pigments as astaxanthin from seaweeds and important bioactive components such as fucoidan, agar, carrageenan, alginate, laminarans, chitin, chitosan, among others which can provide health benefits acting as antimicrobial, anticoagulant, and antioxidant; helping to prevent or lessen symptoms of certain diseases by means of lowering blood glucose levels and strengthen the immune system. Obtaining these components have been achieved thanks to novel technologies and biotechnological advances that have been developed to improve the extraction processes, by the creation of less aggressive processes and methods for both, the feedstock and the environment; besides costs reduction. The aim of this overview is to present the importance of the existence of these nutritional compounds present in marine biomass, its different applications mainly in the food industry and their biological properties that contribute to be consumer.

10.1 INTRODUCTION

Oceans, seas, and coastal areas provide a vital source of nutrition ensuring safe food around the planet. The consumption of products from the sea, such as fish, shrimp, seaweed, shellfish, crabs, jellyfish, sponges, squid, to name a few ones are significant for the rich content of proteins, lipids, carbohydrates, vitamins, and minerals [1]. For these reasons, the aquaculture industry, in recent years has rapidly grown their worldwide productions in developed and first world countries; improving its economies. In 2014, the Food and Agriculture Organization (FAO) reports a fish harvested from aquaculture amounted to 73.8 million tonnes, with an estimated first-sale value of US$160.2 billion [2–4].

Nowadays, the use of fish waste as scales, bones, heads, skin, has been increased for many applications, principally, since 2010, 36% of the fishmeal total feedstock production was obtained from marine food waste (MFW). Previously, the MFW was considered a useless biomass without commercial value; from an economic and nutritional perspective. However, different studies have been made to find applications for these MFW materials for human consumption and other applications in environmental areas with important social and economical impact [2].

Seaweed (also called macroalgae) also fulfills an important role in food area because they are characterized by high biodiversity of species with great variety of nutritional and bioactive compounds, considered by many as the main aquatic food chain, because it chemical composition is constituted by selective polysaccharides, sugars, lipids, vitamins, and minerals that can be applied in nutritional areas [5]. Seaweed industry has an annual global value of USD 5.5–6 billion, mainly used for food, phycocolloids, fertilizer, animal feed additives, cosmetics, and medicines. Products from seaweeds have become very attractive for the food industry; especially for its high capacities to act as thickening and gelling agents that help to improve rheological characteristics of certain products; moreover, it have been reported as sources to provide health benefits acting as anticoagulants, antioxidants, antitumor, antimicrobials, etc. Moreover, it is important to mention that an area under development with an increased interest is the use of seaweeds for bio-alcohols production with significant and positive results, the process is based principally in the biorefinery concept in order to produce biofuels and high added value compounds [4, 6–9].

Additionally, other marine biomass feedstocks with great properties are the residues obtained from shrimp and shellfish production. Shrimp consumption generated 15% of the total sales of the industry marketed fish products in 2010, with an increasing demand every year; [10] but only 65% of shrimp is edible and the rest is waste generated in the production process including shell, head, and tail. These residues have been reported that are rich in high-quality protein, high polyunsaturated fatty acids as omega–3, minerals, carotenoids, chitin, and astaxanthin [11, 12]. Concerning to marine shellfish, the generated residues are a great substrate for the production of biofunctional peptides, carotenoids, and astaxanthin [13]. Crab shells are also a great source of chitin and chitosan used in food, beverages, and water purification while oyster shells and shellfish has been used to produce calcium carbonate and calcium oxide.

The present overview is focused to describe a scenery of the relevant products obtained from sub-valorized marine biomass: seaweeds and others seafood residues; including chemical, biological, and nutritional characteristics, importance, and possible uses to be applied in food industry for direct human consumption or as additives as food ingredients or fortifiers with health benefits (Figure 10.1).

10.2 NUTRIENT EXTRACTS FROM MARINE BIOMASS: SEAWEEDS AND RESIDUES OF FISHING INDUSTRY

10.2.1 SEAWEED

Seaweeds are similar organisms with a simple autotrophic structure, with little or no cell differentiation of complex tissues; there are commonly found attached to rocks or other hard substrates in coastal areas and its classification is based on different properties as their type of pigmentation chemical nature, mechanism of photosynthesis, morphological characteristics, among others [14, 15]. According to FAO statistics in 2014 [3], world production of seaweed come from two sources: harvesting from wild stocks and from aquaculture, the bulk of worldwide production reached of 24.9 million tons in 2012, valued at about USD $6 billion, and 96% (23.8 million tons) of the total production were obtained from aquaculture

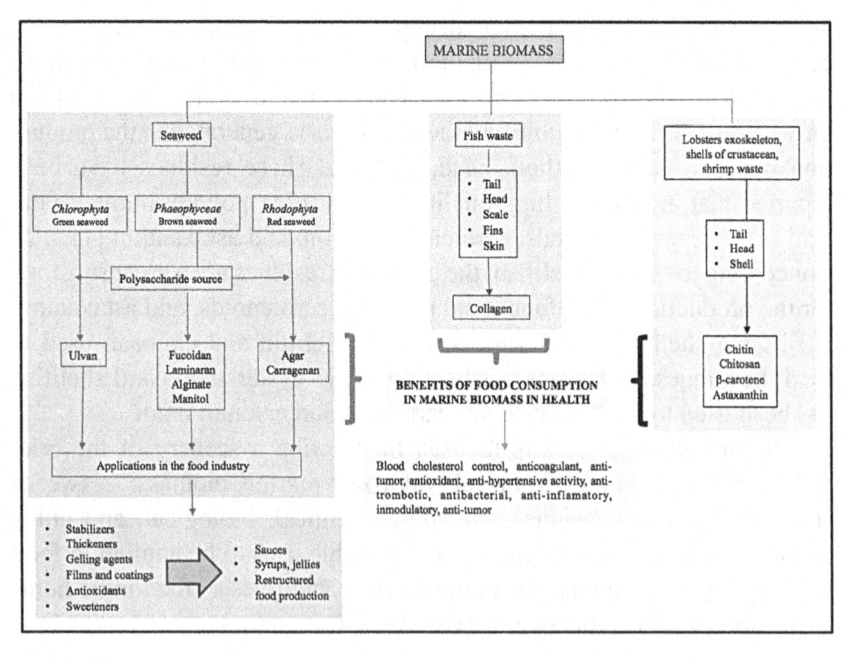

FIGURE 10.1 Main marine biomasses, functional components and applications.

dominated principally by the Asian countries with China, Philippines, and Indonesia as the principals suppliers. On the other hand, Chile, is the top producer of harvesting seaweed from wild stocks, who has remained in first place over the last 10 years, with a regular recovery of 436,035 tons; followed by China and Japan. The European continent has also increased the demand in consumption of seaweed; the main producers are Norway and France with 140,336 and 41,229 tons, respectively. China is the largest producer of edible seaweed with about 5 million tons, mainly *Laminaria japonica,* followed by the Republic of Korea with 800,000 tons, in which 50% is *Undaria pinnatifida*, while Japanese production is around 600,000 tons where 75% corresponds to *Porphyra* species [6, 14–16].

10.2.1.1 POLYSACCHARIDES OF SEAWEEDS

In general, all the seaweed varieties are high in carbohydrates with yields reaching values around 50% dry weight, but this composition varies according to the species to which they belong, geographical location and the water temperature. The cell wall of seaweed contains mostly carbohydrates such as polysaccharides as structure of reserve. There are different types of polysaccharides depending of their classification: (a) brown principally contains alginates, manitol, fucoidan, laminaran; (b) red seaweed present agar and carrageenan, and (c) green seaweed has polydisperse heteropolysaccharide such as glucuronoxylorhamnans, xyloarabinogalactans, and glucuronoxylorhamnogalactans (Figure 10.2A) [6, 17–22].

10.2.1.2 LIPIDS AND PROTEINS

Different types of fatty acids can be found in seaweed ranging from 11–18% of polyunsaturated fatty acids (PUFAs), 31–47% of saturated fatty acids (SFA) and 23–33% of monounsaturated fatty acids (MUFAs) from total fatty acids [23, 24]. Mohamed et al., [23] reported that red and green seaweed types are high in protein content with 10–47% dry weight, while brown seaweed have 5–24% dry weight. The season of collection also affects proteins content, if the collection of the *Rhodophytes* is done during the summer the values are in the range of 12–15% dry weight and in winter, these values decrease.

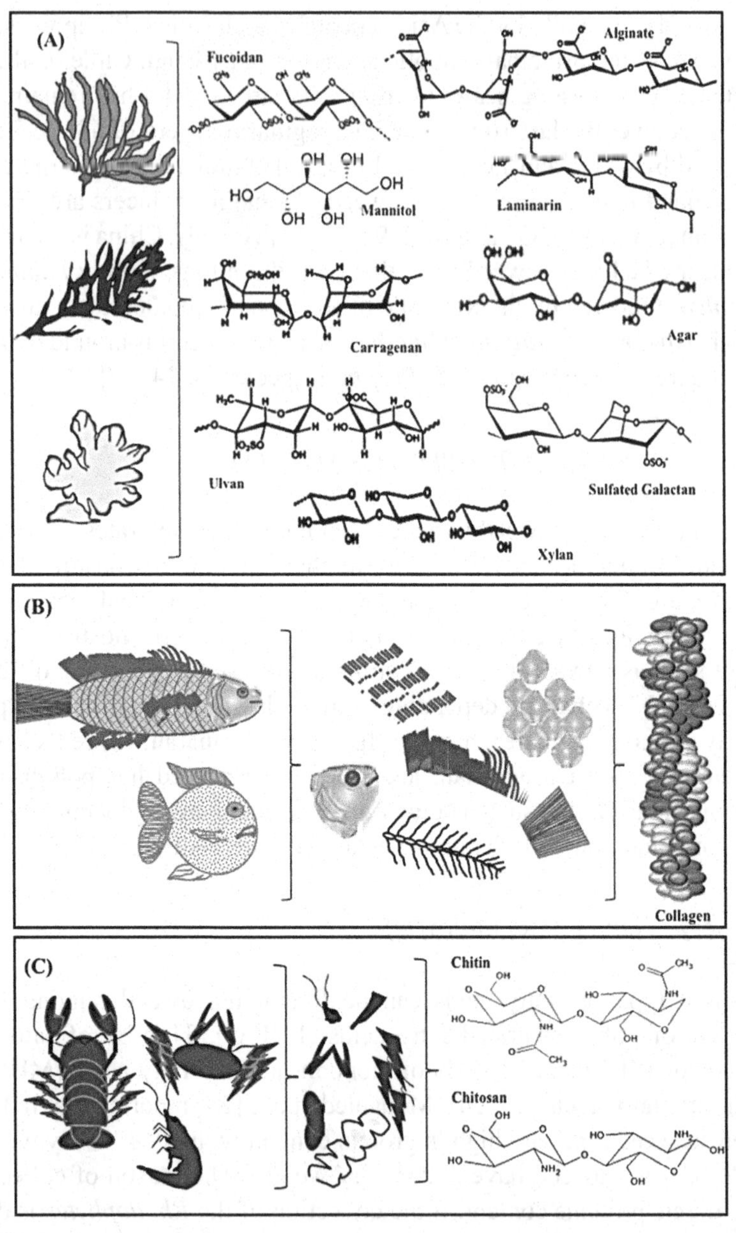

FIGURE 10.2 Principal high value-added compounds applied in food industry extracted from: (a) Brown, red, and green seaweed; (b) Fish; and (c) Crustacean. Adapted from Cervantes-Cisneros et al., 2017 [6].

10.2.1.3 VITAMINS, MINERALS, AND PIGMENTS

Rodrigues et al., [21] determined the most representative seaweeds micro-elements as Mg, Ca, Na, P, and K with 97% of the total mineral content, following by minor content of Al, I, Mn, Fe, Zn, and B. The most significant species with relevance application for its mineral content are *G. gracilis, O. pinnatifida, G. turuturu, S. muticum, S. polyschides* and *C. tomentosum*. The vitamins that can regularly be found are C and α-tocopherol type; also, brown seaweed contains carotenoids like β-carotene, fucoxanthin, violaxanthin, and pheophytins [23, 25].

10.2.2 FISH SCALE, SKIN, BONES, AND FINS

In some countries, the main income is acquired by fish production caused by the considerable increase of the marine products consumption in the last years around the world. FAO [10] estimates in 2014 that the production of aquaculture harvested fish were around 73.8 million tonnes, corresponding to 49.8 million tonnes of finfishes. In 2012, only 60–70% was exploited for human consumption because during the processing of fish in the aquaculture industry, wastes formed by skin, heads, tails, fins, spines, scales, and guts, may generate up to 30–40%, depending on the type of fish, and these wastes are considered low economic value biomass for human consumption. Some of the uses that have been given to these by-products are as fish oil, fishmeal, fish flour, bait, fish for ornamental purposes, pharmaceutical or as animal feed fertilizer in aquaculture, farm animals and pets [26–34]. Actually, there are a variety of techniques developed to apply for recovery different marine biomass nutrients, hence it has been proven that products from these by-products are liable to be used in food industry as additives, while helping the environment by reducing the amount of waste and to generate less economic losses. Moreover, several studies have reported that marine products contain important bioactive compounds that can be used in human consumption because they provide health benefits [26].

Fish waste have large amounts of proteins, fats, and minerals; is important to note that there may be some variation factors in fish waste chemical composition depending of specie, age, season of capture, sex, health, among others [30, 35].

10.2.2.1 PROTEIN

Fish is rich in protein and amino acids, muscle protein is highly nutritious and digestible, it can be found in fish frame, reaching values up to 25% of dry weight, containing 8 essential amino acids. Moreover, another way to obtain better use of protein from fish waste has been its conversion into bioactive peptides and fish protein hydrolysates, the second one is referred to small fragments resulting from degradation by enzymatic hydrolysis; the results are the recovery of 2–20 amino acids with health benefits, ensuring safe consumption by humans and animals feed [26, 30, 36]. One of the main marine source of proteins is collagen, that can be found in invertebrates sources especially from fish skins, fins, scales, and waste material from sponges, jellyfish, and mollusks (Figure 10.2B). Collagen is widely used in cosmetic, pharmaceutical, and food industry because it has the ability to form gelatin in its denatured form and for its thermal stability attributed to the presence of essential and non-essential amino acids like hydroxyproline (produced by hydroxylation of the amino acid proline) [31–38]. Sionkowska et al. [39] reported that there are 28 types of collagen, but three are the principal ones: type I, formed by bones, skin, and tendons; type II composed by cartilage; and type III obtained from skin and viscera. Depending on the type of collagen is the use that is given; the most common classification with several applications is type I, for example, in pharmaceutical industry and medical have been widely used in tissue bioengineering for the regeneration of ligaments, heart valves, blood vessels, scaffolds, in wound healing resulting from burns and ulcers [40–43]. On the other hand, the collagen is used as vehicles for transportation in drug delivery such as capsules, tablet coatings or minipellets because these materials have been shown to be highly biocompatible with the organism due of their high biodegradability [44]. Nowadays, the use of marine collagen in the food industry has increased due to its excellent film-forming capacity, mainly applied as edible biodegradable coatings on fruits, white and red meat, hams, bacon, and sausages, among others. The main advantage of collagen films is as barrier membrane to protect against the migration of oxygen, moistures, and solutes, antimicrobial protection, rancidity decreases, reduce the formation of gases and vapors avoiding food degradation; also, due to the amino acids presence in fish collagen, it can provide antioxidants effect on product directly [41, 45, 46]. Moreover, the combination of marine collagen with other compounds such as chitosan, starch, cellulose, lactic acid, glyceraldehyde, and hydroxyapatite improves their physical and mechanical properties; such as increasing water holding capacity, better texture and stability to increase food quality and safety for its consumption [41, 42, 47].

10.2.2.2 LIPIDS AND OILS

Coppes [48] reported that the amount of oil in fish depends on the species with content values varied around 2–30%. Compared with low-fat content, that can be found in terrestrial animals, the saturated fatty acids (SFA) are the main type of fat that are found in fish and fish wastes, with palmitic acid (16:0) as the most prominent one. Additionally, for monounsaturated fatty acids (MUFA) the principal type is the oleic acid (18:1 omega–9). However, the demand in the consumption of oils from marine species has increased due to the huge interest in two types of n–3 polyunsaturated fatty acids (PUFAs), namely eicosapentaenoic acid (EPA, C20:5 omega–3) and decosahexaenoic acid (DHA, 22:6 omega–3), better known as omega–3 fatty acids, because these fats can only be obtained from seafoods or by taking supplements. The health relevance of these PUFAs is known to have variety of health benefits against cardiovascular diseases including hypotriglyceridemic and anti-inflammatory effects. Also, various studies indicate promising antihypertensive, anticancer, antioxidant, antidepression, antiaging, and antiarthritic effects; and recent studies also indicate insulin-sensitizing effects in metabolic disorders [27, 30, 49].

10.3 WASTE PRODUCTS FROM THE FISH INDUSTRY: MOLLUSC, JELLYFISH, SHRIMP, AND OTHERS

The marine industry focused their production and processing of seafood on mollusk, jellyfish, shrimp, octopus, squid, and cuttlefish, generating a highly significant amount of residues that are primarily composed of heads, tails, shells, legs [50]. Thirty-five countries are the major producers of fish, mollusk, jellyfish, and shrimp, where China occupy the highest position followed by India, Vietnam, Bangladesh, and Egypt with 16.1 million tonnes of mollusks, 7.3 million tonnes of other aquatic animals and 6.9 million tonnes of crustaceans. In 2010, the cephalopods trade reached 4% worldwide of total sales of fish, while shrimp won 15% of sales that increased in 2012 [10]. In the process of shrimp production only a maximum of 65% is edible so this industry approximately generate around 40–50% of the total weight of shrimp waste as head, tail, shell, cephalothorax, and exoskeleton which results in major environmental problems and currently it is required industries to give beneficial use to such waste [1, 11, 12]. The regular uses of seafood residues are as milk substitutes, baked goods, soups, and infant formulas and also enhances food stability [30, 51].

10.3.1 CHITIN AND CHITOSAN

Lobsters exoskeleton, shells of crustacean (crab, mussel shells, prawn shells), shrimp shells, in the cell walls of certain seaweeds and fungi, are the main sources of chitin (15–50% dry weight). Chitin is the second most abundant polysaccharide after cellulose [1, 52–55]. In 1884, it was first identified, as a nontoxic natural polymer with different advantages like high biocompatible with the human-organism, highly biodegradable, low immunity and allergenicity, allowing to be applied in medical, pharmaceutical, and food area [56–60].

The chitin (Figure 10.2C) has high molecular weight, due to intermolecular hydrogen bonds, this polysaccharide is formed by a linear structure with units of (1–4)-linked 2-acetamido–2-deoxy-β-D-glucopyranose. It is insoluble in water, has low reactivity and is less soluble in acidic solvents [1, 50, 61–64]. Anitha et al. [63] reported that chitin can be found in three polymorphic forms in nature, that correspond to α- ways obtained from the crabs and shrimps, its β- found in squids and γ- in loligo; for this reason, its application is very versatile, principally used in the manufacture of nanofibers, sponges, microparticles, delivery drugs. Moreover, is a useful biomaterial for wound healing materials as bandages and dental pieces, in the formation of hydrogels applied in tissue engineering scaffolds, an injectable polymeric hydrogel. In the organism acts as antacids, antiulcer, blood anticoagulant, antitumoral and in the treatment of obesity as nutraceutical/supplement generating satiety sensation and lowering serum cholesterol [56–66]. Edible coatings are the main use in the food industry, ensuring food quality, moreover, other applications are as stabilizing, thickening, and emulsifying, a chelating agent, conservative, and additive of certain foods, as a clarifier and deacidification of juices [56, 66–69].

Chitosan is the product of partial deacetylation of chitin and in contrast to chitin, chitosan is a linear polysaccharide with units of (1–4)-linked 2-amino–2-deoxy-β-D-glucopyranose (Figure 10.2C) [54, 62]. The general method used for obtaining chitin and chitosan, independently of the raw material source, is carried out in three steps. The first stage is the deproteinization and depigmentation, carried out by the removal of proteins with the addition of NaOH solution heated to 60°C following by the depigmentation within acetone for 24 hours. The second step is demineralization, placing the raw material in 1% HCl solution for 4 hours, leaving rest after 24 hours to remove the greater amount of minerals, mainly calcium carbonate. The third and final step is the deacetylation, that consists of heating demineralized chitin in NaOH solution between 100 and 110°C for 6 hours [23, 50, 54, 61, 70]. However, there are biological methods for the extraction of chitin and chitosan. Arbia et al. [52] reported a method with crustacean shells using lactic acid bacteria; these bacteria have the ability to produce shell demineralization, because the lactic acid reacts with calcium. Mohamed et al. [53] report biological extraction methods consisting of enzymatic and microbial fermentation reactions, but these methods have low yields compared with the common extraction method; also other extracted products classification are chitosan oligosaccharides (chito-oligosaccharides, COS) and glucosamine [28, 71].

10.3.2 PROTEINS, AMINO ACIDS, AND LIPIDS

Shrimp wastes are rich in proteins; heads and shells have proteins values of approximately 47% of dry weight, containing important amino acids as isoleucine and lysine and alkaline forms as threonine, lysine, and leucine. Fatty acids in shrimp residues depend on the water temperature of the species collected area with maximum values of 34% of dry weight, including saturated fatty acids followed by acid mono- and poly-unsaturated compounds. In the cephalothorax and exoskeleton (shell) can be found in values of 10.5% and 3.78% dry weight of lipids, whereas the head shows values above 6.9% [1, 11, 12].

10.3.3 VITAMINS, MINERALS, AND PIGMENTS

Mármol et al. [50] mentioned that the main mineral content found in shrimp residues are Mg, N, P, Ca (30–50%) and K with traces presence of Zn, Fe, Ni, and Mn; while in jellyfish source, the principal elements are in a greater proportion of Ca, Cl, Mg, P, K, Na, and Zn [1, 11]. Shahidi et al. [28] reported that is possible the recovery of 10% of carotenoids from the shells and skins; being the main type of β-carotenoid followed by astaxanthin, canthaxanthin, and echinenone, which have been investigated because they exhibit the greatest amount of pigments in comparison with other carotenoids, besides having biological properties as anticancer and antioxidant; in addition acting as precursors of vitamin A.

10.4 BIOACTIVE COMPOUNDS FROM MARINE BIOMASS

There are different techniques for extracting polysaccharides present in seaweeds; the variety of methods include the use of hot water, use of dilute acids and alkalis, solvents, and bases; requiring large amounts of solvents with long periods of extraction [72]. On the other hand, extraction methods through hydrothermal processes present advantages over the methods mentioned above, as higher extraction yield, shorter process extraction time, the maintenance of target compounds bioactivity and the low environment impact [6, 8]. Table 10.1 showed some of the most relevant methods applied for seaweed polysaccharides extraction. Polysaccharides extracted from seaweed, have become very attractive to the industrial sector especially for its rheological characteristics, the most commercial compounds are carrageenan, alginate, and agar because they have important properties such as their ability to form gels, such as stabilizers, thickeners, and emulsifiers, allowing them to be widely used in food industry [7, 8, 15, 73] (Table 10.2). Cervantes-Cisneros et al. [6] reported that the bioactive compound extracted from brown seaweed are mainly fucoidan, alginate, and laminarin; fucoidan and laminarin have been reported as source of different biological activities as antitumor, anti-apoptotic, anti-inflammatory, anticoagulant, and antioxidant activity. In red seaweed the main polysaccharides present are carrageenan (chains with sulfate half-esters attached to the sugar units; of kappa, lambda or iota general forms) and agar (units of agarobiose alternating (1, 3)-linked-D-galactose and (1,

TABLE 10.1 Methods for Seaweed Polysaccharides Extraction

Polysaccharide	Seaweed species	Extraction method			Reference
		Extracted reagent	**Temperature (°C)**	**Other conditions**	
Semi-refined carrageenan	*Eucheuma*	KOH concentration 5%; temperature.	70–80		[73]
	Gigartina	Alkali treatment with ethanol: water – KOH 5%	70–80		[73]
Alginate	*Laminaria digitata*	Water extraction with 1 /16.6 ratio of seaweed/water; sequential heat; addition of sodium carbonate pH 10	80 for 2 h of stirring.	Jacket tank with steam supply (3000 to 4000 mPa.s)	[74]
Fucoidan	*Eisenia bicyclis*	Acid extraction: 0.1 M HCl at for 2 h with twice sequential extraction	60		[17]
	Hydroclathrus clathratus	Distilled water, 100 g in 1 L	23–25 with continuous stirring for 4 h.		[75]
	Fucus vesiculosus	Water, relation seaweed/water of 1:25		125 psi and 1 min using a microwave extractive system.	[72, 76]
			180 for 20 min	Extraction in a steel reactor with oil bath heating by conduction-convection	[76, 77]
	Macrocystis pyrifera	Distilled water, 1 kg of dried and grounded seaweed	55 continuous stirring for 4 h		[78]

TABLE 10.1 *(Continued)*

Polysaccharide	Seaweed species	Extraction method			Reference
		Extracted reagent	**Temperature (°C)**	**Other conditions**	
Laminaran	*Laminaria japonica*	Water as solvent with seaweed/ water ratio of 1:50	50 to 90 for 1 h		[5]
	Sargassum sp.	HCl, 1 g of dried seaweed with 10 and 20 mL of acid	30, 60 and 90, with reaction times of 1.3 to 5 h	Continuous shaking at 200 rpm in a water bath	[79]
Agar	*Gelidium sp.* & *Gracilaria sp.*	Acidified water to pH 6.3–6.5	Boiled temperature		[73]
	Gracilaria verrucosa	Sulfuric acid at 3, 5, 7, and 10% in a total volume of 4000 mL; final pH 6.3–6.6 with in 6 L of water.	Boiled temperature		[80]
K-Carragenan	*Chondracantus chamissoi*	Potassium hydroxide (6%)	Room temperature	Vigorous stirring for 24 hours	[81]

4)-linked-(3, 6)-anhydro-L-galacose residues); about bioactivity, carrageenans had showed anti-tumor and ant-viral properties. Green seaweed is principally constituted by ulvans that are formed by a central backbone of an L-rhamnose 3-sulphated linked to: (a) D-guluronic acid residue; (b) L-iduronic acid residue; (c) D-xylose 4-sulphate residue; or (d) D-xylose residue disaccharide units [5, 18, 83–85].

As was previously mentioned, the main source for obtaining chitin and chitosan comes from crustaceans' discards. These polysaccharides are principally employed in the food technology as films and coatings to increase food shelf life like biodegradable an edible packed to prevent bacteria, fungi, and yeasts growth [35, 86]. Chitin and chitosan are widely used as additive in food industry due to certain mechanical properties such

TABLE 10.2 Most Relevant Applications of Seaweed Polysaccharides in Food Industry

Polysaccharide	Property	Application in the food industry	Reference
Alginate	Stabilizers	Manufacture of ice cream	[23, 74]
	Thickeners	Sauces, syrups, toppings for ice cream and pie fillings	
		Emulsifiers in water/oil solutions for mayonnaise and salad dressings preparation	
	Mixtures of calcium salts and sodium alginate	Supports the formation of gel retarder	
		Jellies, instant desserts	
	Gel	Food production restructured as meats, chicken nuggets, meat pies, fish fillet restructured	
	Films and coatings	Conservation of different food matrix	
	Others	Dietary fiber	
		Beer foam stabilizers	
		Baked goods	
Agar	Gel	Japanese cuisine, in meat and fish products to mimic gelatin and other jelly products.	[7, 82]
Carragenans	Gelling agents, thickeners, and stabilizer	Jellies	
Laminaran	Dietary fiber	As an additive	

as good flexibility, high strength and high durability and its high ability to act as emulsifiers, thickeners, and gelling agents; also, helps to delay the onset of bad odors in the flesh and as clarifiers in beverage as water, apple and carrot juice, and wine production, without affecting the beverage color [50].

Fish wastes as biomass are relevant source to produce bioactive compounds principally rich in proteins, lipids, and minerals. Due to this in the recent decades, fish residues have been regarded as source of nutraceutical ingredients, used to manufacture functional foods based on protein concentrate products as fish oil, rich in unsaturated fatty acids. Additionally, collagen and fish protein hydrolysates (FPH) extracted from fish skeleton has being applied as gelling, emulsification, texture, and whipping, in order to enhance features such as consistency, stability, and elasticity of the product; plus one of the most important applications are in the manufacture of protective films to keep food aroma [36, 87, 88]. The shellfish enzymes have been used as rennet substitute in cheese manufacturing with fish gastric enzymes, fish descaling and the expensive extraction of shellfish proteins [28]. Shrimp shells and heads have been used for the manufacture of sausages protective plastic and as emulsifier and stabilizer [11].

10.5 HEALTH BENEFITS: CONSUMPTION OF FOODS PREPARED WITH MARINE BIOMASS

Buono et al. [89] mentioned that the definition of functional food proposed by the Food and Drug Administration (FDA) "released statements about the relationship between the dietary intake of some foods or nutrients and the prevention of several diseases."

The marine biomass has shown great medicinal and nutraceutical potential like seaweeds compounds such phloratannins, vitamins as A, B_2, B_6, C, and E, dietary fiber and polysaccharides as fucoidan, laminaran, carrageenan, and alginate. These bioactive compounds have been studied against cancer, obesity, diabetes, oxidative stress, inflammation, discomfort and pain, hypertension, and allergies. Due to its medicinal properties, seaweeds have been used for decades as tea and cough medicine to combat colds, bronchitis, and chronic coughs; also as an anticoagulant in blood products and for bowel problems as diarrhea, constipation, and dysentery [18, 84, 90, 91]. Additionally, waste products from the fishery industry (fish wastes, shell of shrimp, mollusc, jellyfish, crab, prawn, krill, and lobster) are rich in bioactive compounds such as, aminoacids, carotenoids as astaxanthin, chitin, and oligosaccharides, also have presence of omega–3 and omega–6 as the most remarkable biocompounds because its acts has anti-inflammatory and blood reducer of lipid content [12, 51, 84, 92, 93]. Kandra et al. [1] reported that chitin most relevant functions are as skin antibacterial, platelet activation, antioxidant, blood cholesterol control and macrophages activation to help immune system for tissues regeneration, also the seaweed pigments as astanxanthin prevents prostate cancer and their pigments synthesize vitamin A (Table 10.3).

10.6 FUTURE PERSPECTIVE

Through years of research, there has been a delay in progress in biotechnological advances with the objective of using marine biomass as a whole: however, in the last decade there has been an increase in researches based on the extraction processes to recovery different bioactive and functional components specifically, highlighting the decrease in extraction times, and obtaining higher yields. Recently, new methods have been studied based to be less aggressive for both, the marine biomass (raw material) and the environment, besides having the advantage of achieving low economic cost. Due to these improvements, researchers have continued searching for novel marine compounds beneficial for health, in order to make possible their benefits on human well-being. However, a critical limitation is that there are also few studies to establish the physiological improvements in humans, so this is an opportunity for researchers interested in performing such studies in order to increase the number of products based on marine biomass in the market of the countries not used to eat these products. Soon, the demand for these compounds will rise considerably, because of the actual and constant increasing tendency to maintain good health through a proper diet. Moreover, it has been demonstrated through the Asian countries that to include products made of raw and extracted components of marine biomass and residues on their daily feed have given relevant health benefits. The significant interest to introduce these products to different parts of the world, are influenced positively, and this has made possible the application and development of new technologies to implement the cultivation of different marine products and the look for applications to the

TABLE 10.3 Health Benefits of Functional Food Ingredient From Different Marine Biomasses

Functional food ingredient	Source	Health benefits	Reference
Chitin Chitosan and its oligosaccharides	Lobsters exoskeleton, the crab and shrimp shells	- Neuroprotective - Antitumoral	[64]
		- Biodegradable, biocompatible, and bioabsorbable compound to be accepted by human body.	[62]
		- Antibacterial with the ability to activate the immune system to prevent invasion of pathogens.	[71]
		- Blood cholesterol control - Antimicrobial activity - Antioxidant application for gene delivery	[1]
		- Anticarcinogenic - Prevention and recovery of renal disease - Reduction of total and LDL cholesterol level	[28]
		- Antilipidemic - Membrane stabilizing properties - Helps increase HDL	[94]
Carotenoides	Waste shrimp and seaweed	- Anti-cancer effect - Prevents cardiovascular and neurodegenerative diseases. - Antioxidant - Anti-inflammatory - Immunomodulatory	[15, 28]
Fucoxanthin	Brown seaweed	- Acts as photoresist and prevents osteoporosis.	[15]
		- Control of metabolic disorders, allowing fight against obesity especially in obese patients with the non-alcoholic fatty liver disease. - Improves insulin resistance and simultaneously reduce blood glucose levels, acts as an anticancer agent.	[23, 28]

TABLE 10.3 *(Continued)*

Functional food ingredient	Source	Health benefits	Reference
Astaxanthin	Red seaweed, lobsters exoskeleton, shrimp, crab, and crawfish shells.	- With vitamin A provides benefits to human health inhibiting bladder and prostate cancer. - Modulates the immune response against tumors cells, reported as ten times more effective in their B-carotene antioxidant activity.	[1]
Ω–3 (EPA,DPA,DHA)	Fish waste and seaweed	- Decrease blood pressure - Antiarrhythmic capacity - Antiinflammatory capacity	[12]
		- Prevents diabetes and other inflammatory diseases - Anti-cancer effect - Essential compounds for proper development and growth of the retina and brain.	[23, 28, 95]
Protein	Waste fish, shrimp, seaweed (green and red)	- Anticoagulant - Anti-tumor - Antioxidant - Antihypertensive activity - Antithrombotic - Antibacterial - Antiinflammatory - In modulator effect	[28]
		- Applied in gastric ulcers treatment	[21, 23]
Minerals as calcium, magnesium, phosphorus, sodium, and potassium	Seaweeds	- Associated with rhinitis decrease in women.	[23]
		- Antioxidant - Anti-cancer - Anti-goiter	[94]
		- Improve mineral content by reducing the salt content	[96]

TABLE 10.3 *(Continued)*

Functional food ingredient	Source	Health benefits	Reference
Dietary fiber	Seaweeds	- Improve slow digestion and caloric intake - Delay gastric emptying moderating appetite to combat obesity - Help in preventing diabetes because it slows the absorption of glucose - Reduce blood cholesterol and functions as prebiotic source.	[97]
Phenolics/ carbohydrates	Seaweeds	- Antioxidant - Antitumor - Antidiabetic - Antibacterial - Anti-inflammatory - Antiallergic - Anticoagulant - Protease inhibitory activities - Cardioprotective.	[28]
		- Antiproliferative - Antiviral	[90]
		- Reduce the risk of breast cancer	[98]
Fucoidan	Brown seaweed	- Anticoagulant properties as neuroprotective cerebral ischemia and Alzheimer - Anti-inflammatory effect - Protects the myocardial membrane	[23, 75, 94]
		- Antitumor - Immunomodulatory - Anticancer - Antivirus - Protects gastric mucosa	[79, 99, 100]

TABLE 10.3 *(Continued)*

Functional food ingredient	Source	Health benefits	Reference
Alginates	Brown seaweed	- Aid in wound healing - Tissue regeneration with the addition of collagen - Anti-tumor	[17, 23, 94] [97]
Laminaran	Brown seaweed	- Help to reduce cholesterol absorption in the intestine and blood. - Anti-apototic - Immuno-stimulatory - Good source of dietary fiber - Anticoagulant - Antioxidant - Antiinflammatory - Anti-tumor activity	[94] [97] [5]
Carragenan and agar	Red seaweed	- Dietary fiber - Decrease cholesterol and blood lipids - Prevent heart disease, atherosclerosis, and hyperlipidemia. - Antiviral activity against dengue virus and papillomavirus.	[94] [23]
Ulvan	Green seaweed	- Helps other nutrients to be more easily absorbed and digested in the jejunum - Antitumoral - Antioxidant - Antiviral - Anticoagulant - Inmunoestimulant	[20, 94]

high quantity of residues. Some of these technologies include the search of optimal conditions for seaweeds growth, tanks improvement for fish production, the expansion areas of companies to reuse marine biomass and residues, seeking to improve the techniques to make the marine products and compounds more effective and profitable, plus it will help to generate more jobs with the market introduction of such products, enhancing the economic income in developed countries

ACKNOWLEDGMENTS

The authors gratefully acknowledged the financial support from the PROMEP project/DSA/103.5/14/10442 – UACOAH – PTC–312 of the Secretary of Public Education of México. Financial support is gratefully acknowledged from the Energy Sustainability Fund 2014–05 (CONACYT-SENER), Mexican Centre for Innovation in Bioenergy (Cemie-Bio), Cluster of Bioalcohols (Ref. 249564). We gratefully acknowledge support for this research by the Mexican Science and Technology Council (CONACYT, Mexico) for the infrastructure project – INFR201601 (Ref. 269461) and CB-2015-01 (Ref. 254808). The author Dulce Arguello thanks to the National Council of Science and Technology (CONACYT-Mexico) for Master fellowship grant.

KEYWORDS

- **bioactive compounds**
- **fish waste**
- **functional foods**
- **marine biomass**

REFERENCES

1. Kandra, P., Challa, M. M., & Kalangi P. J. H., (2012). Efficient use of shrimp waste: Present and future trends. *Appl. Microbiol. Biot.*, *93*, 17–29.
2. Food and Agriculture Organization of the United Nations, (FAO). The state of the world fisheries and aquaculture, FAO: Rome, Italy, 2016. (http://www.fao.org/3/a-i5555e.pdf)

3. West, J., Calumpong, H.P., Martin, G., & (2016). Seaweeds. In: Inniss, L., & Simcock A., (Eds.), *The First Global Integrated Marine Assessment World Ocean Assessment I*; United Nations. (http://www.un.org/Depts/los/global_reporting/WOA_RegProcess.htm).
4. Ruíz, H.A., Rodríguez-Jasso R.M., Aguedo, M., & Kadar, S., (2015). Hydrothermal pretreatment of macroalgal biomass for biorefineries. In: Prokop, A., Bajpai, R. K., & Zappi, M.E., (eds.), *Algal Biorefineries, Products and Refinery Design* (Vol. 2, pp. 467–491). Springer, International Publishing Switzerland.
5. Kadam, S. U., Tiwari, B. K., & O'Donnell, C. P., (2015). Extraction, structure and biofunctional activities of laminarin from brown algae. *Int. J. Food Sci. Tech.*, *50*, 24–31.
6. Cervantes-Cisneros, D.E., Arguello-Esparza, D.G., Cabello-Galindo, A., Picazo, B., Aguilar, C.N., Ruiz, H.A., & Rodríguez-Jasso, R. M., (2017). Hydrothermal processes for extraction of macroalgae high value-added compounds. In : Ruiz, H.A., Thomsen, M. H., & Trajano H. L., (eds.), *Hydrothermal Processing in Biorefineries* (pp. 461–481), Springer.
7. Pereira, L., Gheda, S. F., & Ribeiro-claro, P. J. A., (2013). Analysis by vibrational spectroscopy of seaweed polysaccharides with potential use in food, pharmaceutical, and cosmetic industries. *Int. J. Carbohydr. Chem.*, 1–7.
8. Ruiz, H. A., Rodríguez-Jasso, R. M., Fernandes, B. D., Vicente, A. A., & Teixeira, J.A., (2013). Hydrothermal processing, as an alternative for upgrading agriculture residues and marine biomass according to the biorefinery concept: A review. *Renew. Sust. Energ. Rev.*, *21*, 35–51.
9. Vega-Villasante, F. F., Cortés, L. M., Del, C., Zúñiga, M., L. M., Jaime, C. B., Galindo, L. J. et al., (2010). Small-scale culture of tilapia (*Oreochromis niloticus*), alimentary alternative for rural and peri-urban families in Mexico? REDVET-*Revista Electrónica de Veterinaria.*, *11*, 1–15.
10. Food and Agriculture Organization of the United Nations, (FAO), (2014). The state of the world fisheries and aquaculture; FAO: Rome, Italy.
11. El-Beltagy, A.E., & El-Sayed, S. M., (2012). Functional and nutritional characteristics of protein recovered during isolation of chitin from shrimp waste. *Food Bioprod. Process.*, *90*, 633–638.
12. Sánchez-Camargo, A. P., Almeida, M. M. Â., Lopes, B. L. F., & Cabral, F. A., (2011). Proximate composition and extraction of carotenoids and lipids from Brazilian redspotted shrimp waste (*Farfantepenaeus paulensis*). *J. Food Eng.*, *102*, 87–93.
13. Pádraigín, A., & Harnedy, R. J. F. G., (2012). Bioactive peptides from marine processing waste and shellfish: A review. *J. Funct. Foods.*, *4*, 6–24.
14. Kılınç, B., Cirik, S., & Turan, G., (2014). Organic agriculture towards sustainability. *In Tech.*, 735–748.
15. Quitral, V., Morales, C., Sepúlveda, M., & Schwartz, M. M., (2012). Nutritional and health properties of seaweeds and its potential as a functional ingredient. *Rev. Chil. Nutr.*, *39*, 196–202.
16. United Nations (UN), (2012). *Saccharina lattisima*; UN, 1–10.
17. Ermakova, S., Men'shova, R., Vishchuk, O., Kim, S. M., Um, B. H., Isakov, V. et al., (2013). Water-soluble polysaccharides from the brown alga *Eisenia bicyclis*: Structural characteristics and antitumor activity. *Algal Res.*, *2*, 51–58.

18. Holdt, S. L., & Kraan, S., (2011). Bioactive compounds in seaweed: Functional food applications and legislation. *J. Appl. Psychol., 23*, 543–597.
19. Kraan, S., (2012). Algal polysaccharides, novel applications and outlook. In: *Carbohydrates-Comprehensive Studies on Glycobiology and Glycotechnology* (Vol. 22, pp. 489–524), In Tech. Rijeka, Croatia.
20. Peso-Echarri, P., Frontela-Saseta, C., González-Bermúdez, C. A., Ros-Berruezo, G. F., & Martínez-Graciá, C., (2012). Polysaccharides from seaweed as ingredients in marine aquaculture feeding: alginate, carrageenan and ulvan. *Rev. Biol. Mar. Oceanogr., 47*, 373–381.
21. Rodrigues, D., Freitas, A. C., Pereira, L., Rocha-Santos, T. A. P., Vasconcelos, M. W., Roriz, M., & Duarte, A. C., (2015). Chemical composition of red, brown and green macroalgae from Buarcos bay in central west coast of Portugal. *Food Chem., 183*, 197–207.
22. Rodríguez-Jasso, R. M., Mussatto, S. I., Pastrana, L., Aguilar, C. N., & Teixeira, J. A., (2013). Extraction of sulfated polysaccharides by autohydrolysis of brown seaweed *Fucus vesiculosus*. *J. Appl. Phycol., 25*, 31–39.
23. Mohamed, S., Hashim, S. N., & Rahman, H. A., (2012). Seaweeds: A sustainable functional food for complementary and alternative therapy. *Trends Food Sci. Tech., 23*, 83–96.
24. Sánchez-Muniz, F. J., Bocanegra de Juana, A., Bastida, S., & Benedí, J., (2013). Algae and cardiovascular health. *Functional Ingredients from Algae for Foods and Nutraceuticals*, 369–415.
25. Batista, A. P., Gouveia, L., Bandarra, N. M., Franco, J. M., & Raymundo, A., (2013). Comparison of microalgal biomass profiles as novel functional ingredient for food products. *Algal Res., 2*, 164–173.
26. Chalamaiah, M., Dinesh, K. B., Hemalatha, R., & Jyothirmayi, T., (2012). Fish protein hydrolysates. Proximate composition, amino acid composition, antioxidant activities and applications: A review. *Food Chem., 135*, 3020–3038.
27. Carvajal, A. K., Mozuraityte, R., Standal, I. B., Storrø, I., & Aursand, M., (2014). Antioxidants in fish oil production for improved quality. *J. Am. Oil Chem. Soc., 91*, 1611–1621.
28. Shahidi, F., & Ambigaipalan, P., (2015). Novel functional food ingredients from marine sources. *Curr. Opin. Food Sci., 2*, 123–129.
29. Vignesh, R., Anbarasi, G., Arulmoorthy, M. P., Mohan, K., Rathiesh, A. C., & Srinivasan, M., (2015). Variations in the nutritional composition of the head and bone flours of tilapia (*Oreochromis mossambicus*) adapted to estuarine and freshwater environments. *J. Microbiol. Biotechnol. Food Sci., 4*, 358–364.
30. Ghaly, A. E., Ramakrishnan, V. V., Brooks, M. S., Budge, S. M., & Dave, D., (2013). Fish processing wastes as a potential source of proteins, amino acids and oils: A critical review. *J. Microb. Biochem. Technol., 5*, 107–129.
31. Matmaroh, K., Benjakul, S., Prodpran, T., Encarnacion, A. B., & Kishimura, H., (2011). Characteristics of acid soluble collagen and pepsin soluble collagen from scale of spotted golden goatfish (*Parupeneus heptacanthus*). *Food Chem., 129*, 1179–1186.

32. Olsen, R. L., Toppe, J., & Karunasagar, I., (2014). Challenges and realistic opportunities in the use of by-products from processing of fish and shellfish. *Trends Food Sci. Tech.*, *36*, 144–151.
33. Pati, F., Adhikari, B., & Dhara, S., (2010). Isolation and characterization of fish scale collagen of higher thermal stability. *Bioresource Technol.*, *101*, 3737–3742.
34. Ordóñez-Del Pazo, T., Antelo, L. T., Franco-Uría, A., Pérez-Martín, R. I., Sotelo, C. G., & Alonso, A. A., (2014). Fish discards management in selected Spanish and Portuguese métiers: Identification and potential valorisation. *Trends Food Sci. Tech.*, *36*, 29–43.
35. Jayathilakan, K., Sultana, K., Radhakrishna, K., & Bawa, A. S., (2012). Utilization of byproducts and waste materials from meat, poultry and fish processing industries: A review. *J. Food Sci. Technol.*, *49*, 278–293.
36. Huang, C. Y., Kuo, J. M., Wu, S. J., Tsai, H. T., (2016). Isolation and characterization of fish scale collagen from tilapia (*Oreochromis sp.*) by a novel extrusion–hydro-extraction process. *Food Chem.*, *190*, 997–1006.
37. Ehrlich, H., (2015). Springer. *Biological Materials of Marine Origin*. Biologically-Inspired Systems, *4*, 321–341.
38. Pal, G. K., & Suresh, P. V., (2016). Sustainable valorisation of seafood by-products: Recovery of collagen and development of collagen-based novel functional food ingredients. *Innov. Food Sci. Emerg. Technol.*, *37*, 201–215.
39. Sionkowska, A., Skrzyński, S., Śmiechowski, K., & Kołodziejczak, A., (2017). The review of versatile application of collagen. *Polym. Adv. Technol., 28,* 4–9.
40. Silva, T. H., Moreira-Silva, J., Marques, A. L., Domingues, A., Bayon, Y., & Reis, R. L., (2014). Marine origin collagens and its potential applications. *Mar. Drugs*, *12*, 5881–5901.
41. Jeevithan, E., Qingbo, Z., Bao, B., & Wu, W., (2013). Biomedical and pharmaceutical application of fish collagen and gelatin: A review. *Journal of Nutritional Therapeutics*, *2*, 218–227.
42. Atef, M., Ojagh, S. M., (2017). Health benefits and food applications of bioactive compounds from fish byproducts: A review. *J. Funct. Foods.*, *35*, 673–681.
43. Cao, H., Chen, M.M., Liu, Y., Liu, Y.Y., Huang, Y.Q., Wang, J.H. et al., (2015). Fish collagen-based scaffold containing PLGA microspheres for controlled growth factor delivery in skin tissue engineering. *Colloids Surf. B.*, *136*, 1098–1106.
44. Yamada, S., Yamamoto, K., Ikeda, T., Yanagiguchi, K., & Hayashi, Y., (2014). Potency of fish collagen as a scaffold for regenerative medicine. *BioMed Res. Int.*, *14*, 1–8.
45. Liang, J., Li, Q., Lin, B., Yu, Y., Ding, Y., Dai, X., & Li, Y., (2014). Comparative studies of oral administration of marine collagen peptides from chum salmon (*Oncorhynchus keta*) pre- and post-acute ethanol intoxication in female Sprague-Dawley rats. *Food Funct.*, *5*, 2078–2085.
46. Yamamoto, K., Yoshizawa, Y., Yanagiguchi, K., Ikeda, T., Yamada, S., & Hayashi, Y., (2015). The characterization of fish (tilapia) collagen sponge as a biomaterial. *Int. J. Polym. Sci.*, 1–5.
47. Sionkowska, A., Skrzyński, S., Śmiechowski, K., & Kołodziejczak, A., (2017). The review of versatile application of collagen. *Polym. Adv. Technol.*, *28*, 4–9.

48. Coppes, P. Z., (2015). Chemical composition of fish and fishery products. *Handbook of Food Chemistry*, 403–435.
49. Siriwardhana, N., Kalupahana, N. S., & Moustaid-Moussa, N., (2012). Health benefits of n–3 polyunsaturated fatty acids: Eicosapentaenoic acid and docosahexaenoic acid. Se-Kwon Kim, (ed.), *Advances in Food and Nutrition Research. Marine Medicinal Foods. Implications and Applications: Animals and Microbes*, *65*, 211–222.
50. Mármol, Z., Páez, G., Rincón, M., Araujo, K., & Aiello, C., (2011). Chitin and Chitosan friendly polymer.A review of their applications. *Rev. Tecnocientifica URU*, 53–58.
51. Prameela, K., Venkatesh, K., Immandi, S.B., Kasturi, A.P.K., Krishna, R., Murali, Ch., (2017). Next generation nutraceutical from shrimp waste: The convergence of applications with extraction methods. *Food Chem.*, *237*, 121–132.
52. Arbia, W., (2013). Chitin extraction from crustacean shells using biological methods: A review. *Food Technol. Biotech.*, *51*, 12–25.
53. Mohammed, M. H., Williams, P. A., & Tverezovskaya, O., (2013). Extraction of chitin from prawn shells and conversion to low molecular mass chitosan. *Food Hydrocol.*, *31*, 166–171.
54. Puvvada, Y. S., Vankayalapati, S., & Sukhavasi, S., (2012). Extraction of chitin from chitosan from exoskeleton of shrimp for application in the pharmaceutical industry. *Int. Curr. Phar. J.*, *1*, 258–263.
55. Zhang, X., Geng, X., Jiang, H., Li, J., & Huang, J., (2012). Synthesis and characteristics of chitin and chitosan with the (2-hydroxy–3-trimethylammonium) propyl functionality, and evaluation of their antioxidant activity *in vitro*. *Carbohyd. Polym.*, *89*, 486–491.
56. Usman, A., Zia, K., Zuber, M., Tabasum, S., Rehman, S., & Zia, F., (2016). Chitin and chitosan based polyurethanes: A review of recent advances and prospective biomedical applications. *Int. J. Biol. Macromol.*, *86*, 630–645.
57. Thakur, V. K., & Thakur, M. K., (2014). Recent advances in graft copolymerization and applications of chitosan. A review. ACS Sustain. *Chem. Eng.*, *2*, 2637–2652.
58. Kyzas, G. Z., Bikiaris, D. N., (2015). Recent modifications of chitosan for adsorption applications: A critical and systematic review. *Mar. Drugs.*, *13*, 312–337.
59. Azuma, K., Ifuku, S., Osaki, T., Okamoto, Y., & Minami, S., (2014). Preparation and biomedical applications of chitin and chitosan nanofibers. *J. Biomed. Nanotechnol.*, *10*, 2891–2920.
60. Cheung, R., Ng, T., Wong, J., & Chan, W., (2015). Chitosan: An update on potential biomedical and pharmaceutical applications. *Mar. Drugs.*, *13*, 5156–5186.
61. Abdulkarim, A., Isa, M. T., Abdulsalam, S., Muhammad, A. J., & Ameh, A. O., (2013). Extraction and characterization of chitin and chitosan from mussel shell. *Civil Environ. Res.*, *3*, 108–114.
62. Younes, I., & Rinaudo, M., (2015). Chitin and chitosan preparation from marine sources. Structure, properties and applications. *Mar Drugs*, *13*, 1133–1174.
63. Anitha, A., Sowmya, S., Kumar, P. T. S., Deepthi, S., Chennazhi, K. P., & Ehrlich, H., (2014). Chitin and chitosan in selected biomedical applications. *Prog. Polym. Sci.*, *39*, 1644–1667.

64. Benhabiles, M. S., Salah, R., Lounici, H., Drouiche, N., Goosen, M. F. A., & Mameri, N., (2012). Antibacterial activity of chitin, chitosan and its oligomers prepared from shrimp shell waste. *Food Hydrocol., 29*, 48–56.
65. Croisier, F., & Jérôme, C., (2013). Chitosan-based biomaterials for tissue engineering. *Eur. Polym. J., 49*, 780–792.
66. Nilsen-Nygaard, J., Strand, S. P., Vårum, K. M., Draget, K. I., & Nordgård, C. T., (2015). Chitosan: gels and interfacial properties. *Polym., 7*, 552–579.
67. Lopez, O., Garcia, M., Villar, M. A., Gentili, A. A., Rodriguez, M. S., & Albertengo, L., (2014). Thermo-compression of biodegradable thermoplastic corn starch films containing chitin and chitosan. *LWT— Food Sci. Technol., 57*, 106–115.
68. Abdel-Rahman, R. M., Hrdina, R., Abdel-Mohsen, A. M., Soliman, A. Y., Mohamed, F. K., Mohsin, K. et al., (2015). Chitin and chitosan from Brazilian Atlantic Coast: Isolation, characterization and antibacterial activity. *Int. J. Biol. Macromol., 80*, 107–120.
69. Leceta, I., Guerrero, P., De la Caba, K., (2013). Functional properties of chitosan-based films. *Carbohyd Polym., 93*, 339– 346.
70. Radwan, M. A., Farrag, S. A. A., Abu-Elamayem, M. M., & Ahmed, N. S., (2012). Extraction, characterization, and nematicidal activity of chitin and chitosan derived from shrimp shell wastes. *Biol. Fert. Soils., 48*, 463–468.
71. Limam, Z., Selmi, S., Sadok, S., & Abed, A. E. l., (2011). Extraction and characterization of chitin and chitosan from crustacean by-products: Biological and physicochemical properties. *Afr. J. Biotechnol., 10*, 640–647.
72. Rodriguez-Jasso, R. M., Mussatto, S. I., Pastrana, L., Aguilar, C. N., & Teixeira, J. A., (2011). Microwave-assisted extraction of sulfated polysaccharides (fucoidan) from brown seaweed. *Carbohydr. Polym., 86*, 1137–1144.
73. Hernández-Carmona, G., Freile-Pelegrín, Y., & Hernández-Garibay, E., (2013). Conventional and alternative technologies for the extraction of algal polysaccharides. *Functional Ingredients from Algae for Foods and Nutraceuticals*, 475–516.
74. Romero, G. C. A., Malo, L., & Palou, E., (2013). Properties and food applications of alginate (Propiedades del alginato y aplicaciones en alimentos). *Temas Selectos de Ingeniería de Alimentos., 1*, 87–96.
75. Lozano, I. S., (2013). Extraction, characterization and anticoagulant activity of brown alga *Hydroclathrus clathratus* (C. Agardh) M. A. Howe sulphated polysaccharides. Thesis Autonomous University of Baja California Sur, Mexico.
76. Rodríguez-Jasso, R. M., Mussatto, S. I., Pastrana, L., Aguilar, C. N., & Teixeira, J. A., (2013). Extraction of sulfated polysaccharides by autohydrolysis of brown seaweed *Fucus vesiculosus. J. Appl. Phycol., 25*, 31–39.
77. Rodriguez-Jasso, R. M., Mussatto, S., Pastrana, L., Aguilar, C., & Teixeira, J., (2014). Chemical composition and antioxidant activity of sulphated polysaccharides extracted from *Fucus vesiculosus* using different hydrothermal processes. *Chem. Pap., 68*, 203–209.
78. Silva, E. S. C., (2013). In vivo evaluation of the sulphated activity extracted from the alga *Macrocystis pyrifera*. Thesis. Interdisciplinary Center for Marine Sciences (Centro interdisciplinario de ciencias marinas) – IPN, Mexico.

79. Ale, M. T., & Meyer, A. S., (2013). Fucoidans from brown seaweeds: An update on structures, extraction techniques and use of enzymes as tools for structural elucidation. *RSC Adv.*, *3*, 8131–8141.
80. Montilla-Escudero, E. A., Dulce-Rivedeneira, M. F., Quevedo-Hidalgo, B., Mercado-Reyes, M., Álvarez-León R., Molina-Vargas, J. N. et al., (2011). Efecto del tratamiento alcalino sobre la productividad y las propiedades físicas del agar-agar proveniente de *Gracilaria verrucosa*. *Bol. Invest. Mar.Cost.*, *40*, 75–88.
81. Salas, N., Córdova, C., & Estrada, E., (2008). κ-Carrageenan and λ-Carrageenan from *Chondracanthus chamissoi* macroalgae and its application in the food industry (Obtención de κ-Carragenano y λ-Carragenano a partir de macroalga Chondracanthus chamissoi y su aplicación en la industria alimentaria). *Industrial Data Revista de Investigación.*, *11*, 52–58.
82. Gómez-Ordóñez, E., (2013). Nutritional evaluation and biological properties of edible marine algae. In vitro and in vivo studies (Evaluación nutricional y propiedades biológicas de algas marinas comestibles. Estudios in vitro e in vivo). *Thesis: Universidad Complutense de Madrid*, Spain.
83. Balboa, E. M., Rivas, S., Moure, A., Domínguez, H., & Parajó, J.C., (2013). Simultaneous extraction and depolymerization of fucoidan from *Sargassum muticum* in aqueous media. *Mar. Drugs.*, *11*, 4612–4627.
84. Hamed, I., Özogul, F., Özogul, Y., & Regenstein, J. M., (2015). Marine bioactive compounds and their health benefits: A review. *Compr. Rev. Food Sci. Food. Saf.*, *14*, 446–465.
85. Vera, J., Castro, J., Gonzalez, A., & Moenne, A., (2011). Seaweed polysaccharides and derived oligosaccharides stimulate defense responses and protection against pathogens in plants. *Mar. Drugs*, *9*, 2514–2525.
86. Souza, B. W. S., Cerqueira, M. A., Ruiz, J. T. M., Casariego, A., Teixeira, J. A., & Vicente, A. A., (2010). Effect of chitosan-based coatings on the shelf life of salmon (*Salmo salar*). *J. Agric. Food Chem.*, *58*, 11456–11462.
87. Wang, B., Wang, Y. M., Chi, C. F., Luo, H. Y., Deng, S. G., & Ma, J. Y., (2013). Isolation and characterization of collagen and antioxidant collagen peptides from scales of Croceine croaker (*Pseudosciaena crocea*). *Mar. Drugs*, *11*, 4641–4661.
88. Weng, W., Zheng, H., & Su, W., (2014). Characterization of edible films based on tilapia (*Tilapia zillii*) scale gelatin with different extraction pH. *Food Hydrocol.*, *41*, 19–26.
89. Buono, S., Langellotti, A. L., Martello, A., Rinna, F., & Fogliano, V., (2014). Functional ingredients from microalgae. *Food Funct.*, *5*, 1669–1685.
90. Santos, S. A. O., Vilela, C., Freire, C. S. R., Abreu, M. H., Rocha, S. M., & Silvestre, A. J. D., (2015). *Chlorophyta* and *Rhodophyta* macroalgae Abu-Elamayem a source of health promoting phytochemicals. *Food Chem.*, *183*, 122–128.
91. Cardoso, S. M., Pereira, O. R., Seca, A. M. L., Pinto, D. C. G. A., Silva, A. M. S., (2015). Seaweeds as preventive agents for cardiovascular diseases: From nutrients to functional foods. *Mar. Drugs.*, *13*, 6838–6865.
92. Harnedy, P. A., & FitzGerald, R. J., (2012). Bioactive peptides from marine processing waste and shellfish: A review. *J. Funct. Foods*, *4*, 6–24.

93. Mao, X., Guo, N., Sun, J., & Xue, C., (2017). Comprehensive utilization of shrimp waste based on biotechnological methods, a review. *J. Clean. Prod.*, *143*, 814–823.
94. Mayakrishnan, V., Kannappan, P., Abdullah, N., & Ahmed, A. B. A., (2013). Cardioprotective activity of polysaccharides derived from marine algae: An overview. *Trends Food Sci. Tech.*, *30*, 98–104.
95. Schmid, M., Guihéneuf, F., & Stengel, D. B., (2014). Fatty acid contents and profiles of 16 macroalgae collected from the Irish Coast at two seasons. *J. Appl. Phycol.*, *26*, 451–463.
96. Kılınç, B., Cirik, S., & Turan, G., (2013). Seaweeds for food and industrial applications. *Food Ind.*, 735–748.
97. Gupta, S., & Abu-Ghannam, N., (2011a). Bioactive potential and possible health effects of edible brown seaweeds. *Trends Food Sci. Tech.*, *22*, 315–326.
98. Gupta, S., & Abu-Ghannam, N., (2011b). Recent developments in the application of seaweeds or seaweed extracts as a means for enhancing the safety and quality attributes of foods. *Innov. Food Sci. Emerg.*, *12*, 600–609.
99. Sinurat, E., & Marraskuranto, E., (2012). Fucoidan from brown seaweed and its bioactivity. *Squalen. Bulletin of Marine and Fisheries Postharvest and Biotechnology*, *7*, 131–138.
100. Wijesinghe, W. A. J. P., & Jeon, Y.J., (2012). Biological activities and potential industrial applications of fucose rich sulfated polysaccharides and fucoidans isolated from brown seaweeds. A review. *Carbohyd. Polym.*, *88*, 13–20.

CHAPTER 11

FUNCTIONALITY FEATURES OF CANDELILLA WAX IN EDIBLE COATINGS

OLGA B. ALVAREZ-PEREZ,[1] MIGUEL ÁNGEL DE LEÓN-ZAPATA,[1] ROMEO ROJAS MOLINA,[1] JANETH VENTURA-SOBREVILLA,[1] MIGUEL A. AGUILAR-GONZÁLEZ,[2] and CRISTÓBAL N. AGUILAR[1]

[1] *Group of Bioprocesses and Bioproducts, Food Research Department, School of Chemistry, Universidad Autonoma de Coahuila, 25280 Saltillo, Coahuila, México, E-mail: cristobal.aguilar@uadec.edu.mx*

[2] *Cinvestav-IPN, Unit Saltillo, Coahuila, Mexico*

ABSTRACT

The use of lipids in the development of edible coatings is a viable alternative to conserve natural and fresh products, allowing the extension of the shelf life quality without causing any harm to the environment, compared to films prepared from synthetic plastics. For this reason, the most important lipid in this application is the wax, because it possesses the necessary properties to perform this function. Since its nature, the fruits tend to have a layer of wax that confers protection against pathogens, environmental, pests, etc. In this review, we report the importance of candelilla wax as a natural resource from the Mexican semi-arid region, as well as their uses in the development of edible coatings, because it is a biodegradable and edible material approved by the FDA. Its importance as an essential component of an edible coating made with this type of biomaterial is based in its action as a plasticizer. Also, we describe some reports of the development of nutraceutical edible coatings formulated with the addition of natural phenolic antioxidants.

11.1 INTRODUCTION

The concern for the mass production of waste from packaging materials has meant a decisive shift in the vision towards the use of biodegradable materials. Especially those coming from agricultural surpluses, due to the growing demand by consumers of food made from natural products, has led to the innovation of new conservation technologies at agro-industrial level, which help prolong the shelf life of fresh fruit [1]. However, the use of natural coatings cannot, nor intends to, replace the usage of traditional packaging materials, but it is necessary to take into account its functional characteristics and the possible advantages of behavior in certain applications [2].

The technical challenges involved in producing food and preserving them with stable quality, indicate that the use of this type of coatings will be greater than what it currently is; However, despite the fact that the technical information available for the elaboration of edible covers is wide, it is not universal for all products, which implies a challenge for the development of specific coatings and films for each food [3]. In the particular case of fruits and vegetables for fresh consumption, the edible coatings provide an additional protective cover whose technological impact is equivalent to that of a modified atmosphere, which consists of a thin and continuous layer, made of materials that can be ingested, and provides a barrier to moisture, oxygen, and solutes, this can completely cover the food or can be placed in the components of the product [4] and must ensure the stability of the Food and prolong its useful life.

According to the conditions of storage of fruits and vegetables should be considered some factors whether mechanical or chemical involved in the design of films [4], therefore represent a storage alternative for products that can be consumed in fresh [3]. The edible coatings that are being tested in post-harvest are mixed formulations of lipid compounds and hydrocolloids, whose main priority is the preservation and protection of products of plant origin, against microbial contamination generated during manipulation [5]. The use of Candelilla wax has been reported as a basic component in the elaboration of edible coatings evaluating different factors involved in the conservation of fresh fruits. The Candelilla wax is extracted from the wild plant Euphorbia *Antisyphilitica Zucc.*, it represents one of the main biopolymers that present biodegradable and edible properties, its structure is amorphous and its hardness is of an intermediate degree between that of the wax of carnauba and that of Bee [6].

Candelilla Wax is considered a GRAS substance by the FDA, so it has multiple applications in the food industry [7], that is why it is used for the elaboration of edible roofs, providing one of the best barriers permeable to moisture and gases that are the product of the metabolism of the fruit. Candelilla is a perennial plant that develops in semi-desert climates, almost devoid of leaves. The Candelilla is one of the plants that grow in the wild with a greater number of applications of use. It is reproduced by both aerial and underground stem shoots and seed. The collection of the Candelilla plant for the production of natural wax has been one of the most important economic activities in five States of the Mexican Republic. It is currently being used in more than 20 different industries around the world, mainly in the United States, the European Union, and Japan. Its distinctive properties confer on it the category of raw material essential for the manufacture of cosmetics, inks, paints, adhesives, coatings, brighteners and polishes, electrical insulators, integrated circuits, chewing gums, fruit coatings for export purposes, thinners and hardeners of other waxes, candles against insects, among others [46].

At the pharmaceutical level to the Candelilla is recognized several therapeutic properties. Currently in Mexico there are research projects aimed at technological improvement in the process of extraction and purification of timber and non-timber forest products with high commercial value, in particular, the Department of Food Research of the Faculty of Chemical Sciences of the Autonomous University of Coahuila, has reported the elaboration of edible roofs with the addition of natural antioxidants produced by fungal fermentation in solid medium, they have presented positive results for the conservation of Fresh fruits [8], providing a barrier against pathogenic microorganisms that may possibly damage the fruit [9] and give the consumer a health benefit, thanks to the presence of antioxidants, which have anti-tumor, anticarcinogenic, and anti-cancer properties [8].

11.2 NATURAL WAXES USED IN THE FORMULATION OF EDIBLE COATINGS

The use of waxes to cover the fruits by immersion, is one of the oldest methods, practiced for the first time in China [10], since the early 12th century [11], retarding perspiration in lemons and oranges, as they are more effective in blocking the migration of moisture, specifically during seasonal changes and continues to be used in other types of fruits [10], being the Candelilla one of the most resistant [12].

Natural waxes applied to fresh perishable products to reduce perspiration are: beeswax, carnauba wax, candelilla wax and rice bran wax [12], also paraffin waxes are some of the waxes prepared and used in the elaboration of edible coatings, which are also used as microencapsulation agents., specifically for substances that provide fruit smells and flavors [1]. Edible waxes are significantly more resistant to moisture transport than most other lipid or non-lipid films [12], in addition to preventing the softening caused by enzymatic hydrolysis of plant cells and membrane components during the cutting process [13], however it is important that wax covers in fresh or perishable fruits are not completely waterproof, which causes anaerobes favoring the physiological disorders that shorten the Half-Life [45]. Waxing is a conservation technique widely used by marketers, supermarkets, and exporters in the world, whose method generates a barrier of protection between the product and the environment to prevent the fruit from breathing less or deteriorate faster, this wear is characterized by the loss of moisture or dehydration of horticultural products and is a deterioration factor so we must try to maintain an optimum quality of the product [45].

11.3 CANDELILLA WAX *(EUPHORBIA ANTISTISYPHILIICA ZUCC.)*

The Candelilla wax is extracted from the wild plant Euphorbia *Antisyphilitica Zucc*., which is formed by esters of long-chain fatty acids that create a protective surface in the plant [8]. It is insoluble in water, but highly soluble in acetone, chloroform, benzene, and other organic solvents [6].

It presents a wide variety of applications, being currently used in more than 20 different industries around the world (Table 11.1).

Its distinctive properties confer on the category of essential material for the manufacture of cosmetics, inks, adhesives, coatings, emulsions, polishes, and pharmaceutical products. In the cosmetics industry, it is being a good plasticizer [14]. It is also used in the manufacture of chewing gum, in the smelting, molding industry, in manufacturing various products in the electronic and electrical industries. There are many other applications where it is currently used, including cardboard coatings, crayon manufacturing, paints, wax candles, lubricants, paper coatings, anticorrosives, waterproofing, and Fireworks [14]. Candelilla's wax is recognized by the Food and Drug Administration of the United States of America (FDA), as a natural safe-GRAS substance, generally Recognized as a safe-for application in the food industry, therefore it is widely used in various sectors of the branch [7]. Because Candelilla wax is an edible wax, it is being used for the elaboration of natural coatings that can retard the ripening and ageing of fruits and vegetables, maintaining a controlled atmosphere on the exterior surface, which allows the protection of the product in the face of environmental, transport, and storage conditions [6].

TABLE 11.1 Commercial Importance of Mexican Candelilla Wax [44]

Importing country	Imported candelilla wax (2016) (USD)	Increase in purchase in the last five years (%)
Germany	11,000.00	23.3
China	9,900.00	11.0
Italy	5,600.00	23.6
France	5,200.00	10.0
Spain	3,800.00	27.5

11.4 EFFECT OF CANDELILLA WAX AS PLASTICIZER ON THE FUNCTIONALITY OF AN EDIBLE COATING

Plasticization is a very important factor in the formulation, since they affect the mechanical properties [15] and physical of the coating (elasticity, flexibility, permeability, wettability) [16], because they alter the mobility of the chain, the diffusion coefficients of gas or water and the structure of the films [15], reducing the intermolecular forces between the polymer chains and increasing the free volume [17], consequently, there is more space for water molecules to migrate, as well as hydrophilic plasticizers such as glycerol, are compatible with the polymeric material that forms the film and increase the absorption capacity of polar molecules such as water [18]. The increase in permeability, with the plasticizer content, may be related to the hydrophilicity of the plasticizer molecule [19], because the permeability to the water vapor increases as the plasticizer content increases, however up to 30% glycerol content, that increase is relatively mild, later observed a more pronounced increase [20].

Cellulose-based coatings, they are very efficient barriers to the permeability of oxygen and its property of barrier to the water vapor [21], these can be improved with the addition of lipids as Plasticizers [21], since they generally increase the permeability of the same [17]. The application of a lipid layer on the surface of fruits replaces the natural waxes of the cuticle, which may have been partially removed during washing [12]. The edible wax covers of candelilla, have different functional properties, because when mixed with oils and polymers of high molecular weight as natural gums, they have an effect on the fruit to be coated, avoiding weight loss [20], the use of Candelilla wax on combined edible roofs has been amply evidenced by Bosquez-Molina et al. [1], who demonstrated that the covers with this material and rubber mesquite create a modified atmosphere inside the fruit, to retard the process of maturation and senescence in a way similar to that of a controlled atmosphere that is much more expensive [20], also avoids an increase in the production of ethylene and the hauling of additives that retard the discoloration and microbial growth [2, 22], allow to control the respiration of the product, providing better permeability and texture, since it modifies the mechanical properties; Fulfilling the function of Plasticizer [23], weakening the intermolecular forces between the polymer chains, increasing the flexibility of the coating [20]. The use of a hydrophilic or hydrophobic plasticizer will produce a coating with similar characteristics [5]. Lipids such as Candelilla wax are good plasticizers, so this natural wax product has shown its advantages over most synthetic waxes used in this industry [14], they are compounds of low volatility and function as plasticizers, which are added to the coating [5], considering themselves two forces, one among the forming molecules of the film, called cohesion and another in the coating and substrate, called adhesion [24], in order to reduce the fragility, increasing the flexibility, hardness, and resistance to cutting, as they decrease the intramolecular forces of the polymer chains, thus producing a decrease in the strength of cohesion, tension, and in the vitreous transition temperature.

Candelilla Wax covers, among others, can be used as a support when adding preservatives or other additives on the surface of foodstuffs, mainly fresh fruits and vegetables to prolong periods of post-harvest storage, which consist of an emulsion made of waxes and oils in water, which are sprinkled in the fruits to improve their appearance, brightness, color, softness, control its maturity and retard the loss of water [25]. The emulsion originated in the elaboration of an edible cover based on Candelilla wax must present an adequate homogenization of the system and in this way guarantee the uniformity in the size and distribution of the particles of the dispersed phase [20], as it will be reflected in the final barrier properties such as water vapor permeability and gases [20]. It is important to know the volumetric fraction of the dispersed phase of the wax emulsion of Candelilla, as it influences much on the appearance [2].

11.5 FUNCTIONAL EDIBLE COATINGS BASED ON CANDELILLA WAX SUPPLEMENTED WITH PHENOLIC ANTIOXIDANTS

Additives are used to impart mechanical, nutritional, and organoleptic properties to edible roofs [26], these may be of the plasticizer type (polyhydric alcohols, oils, fatty acids), surfactant, and emulsifier type (fats, oils, polyethylene glycol) chemical preservatives (benzóico acid, sodium benzoate, sorbic acid, potassium sorbate, propionic acid) [2], as well as antimicrobial agents, antioxidants, dyes, flavorings, and calcium as a firming agent of cell membranes, among others [26], can be applied to control and modify surface conditions, reducing some of the degrading reactions [27, 28]. The maintenance of microbial stability can be obtained using edible coatings with antimicrobial action and combined with refrigeration and controlled atmosphere. For fruits, waxes are usually used with the addition of sorbic acid and sorbates as antifungal [27, 28].

The influence of a given additive will depend on its concentration, chemical structure, the degree of dispersion in the film and the degree of interaction with the polymer [5]. Some chemicals and natural products are used as antioxidants or as microbicides; Ascorbic acid, citric, and lemon juice are used as antioxidants and salts of 5-acetyl-8-hydroxyquinoline or strong inorganic acids such as H_2SO_4 or H_3PO_4 are used as microbicides with very good results. The use of these substances does not prevent or retard the maturation so, other methods should be used together for this purpose, such as the application of gamma radiation that modifies the ripening time of the fruit according to its dose; However some defects have been detected, such as the darkening of the pulp becoming coffee or by contrast discoloration. Good results have been reported when radiation is less than 7 J/kg [29].

Zhang et al. [30] reported that the use of ascorbic acid, isoascorbic acid and acetyl cysteine reduced darkening in litchi. Buta et al. [31] used combinations of 4-hexylresorcinol, isoascorbic acid, CaCl2, and acetyl cysteine to reduce changes in apple slices. Luo et al. [32] controlled darkening in apple slices using 4-hexylresorcinol mixed with ascorbic acid. Baldwin et al. [2] observed better protection with the addition of ascorbic acid in edible roofs. Ruiz-Cruz et al. [33] showed the positive effect of different antioxidants (independent and mixed) in inhibiting the darkening of cut fresh pineapple, because when antioxidants are used in combination with other technologies: treatments with heat, modified, and controlled atmospheres, edible covers, gamma radiation, and electromagnetic pulses the darkening in fruits is inhibited.

Saucedo-Pompa et al. [8] reported for the first time the use of gallic acid, ellagic acid and Aloe Vera in the formulation of an edible cover based on Candelilla wax, as anti-darkening additives of fruits, showing excellent results, even the controls with the cover without antioxidants were better compared with the fruits without edible cover [8]. The addition of ellagic acid and aloe vera as natural antioxidants to the edible Candelilla wax cover showed positive results when applied to fresh fruits, this due to the protective barrier that represents the cover of wax of candelilla as physical barrier of the coating, allowing greater control of gases, greater permeability and therefore the better control in the respiration of the fruit and in turn the addition of antioxidants, they intervene inhibiting microbial growth, as well as the possibility of providing a benefit to the health of the consumer, due to the anti-cancer and anti-tumor properties that present this type of antioxidants [8]. Saucedo-Pompa et al. [9] reported that when applying the edible cover based on Candelilla wax with the addition of ellagic acid and aloe vera to freshly cut avocados decreases the aqueous activity of the fruit, these results indicate that for freshly cut fruits reduction in weight loss is very important and the use of edible covers carrying natural antioxidants is an excellent tool to control weight loss [22, 34, 35]. The apple slices with the edible covers with ellagic acid and aloe vera kept to a greater extent the initial firmness, these covers had a protective effect on the firmness of freshly cut fruits. The texture of freshly cut fruits was improved with the application of edible Candelilla wax covers, these results are similar to those reported by Ghaouth et al. [36] who applied edible covers in tomato. Saucedo-Pompa et al. [9], reported antifungal properties of ellagic acid and Aloe Vera as part of the formulation of a candelilla wax-based edible cover which was applied in avocados, its functionality allowed to increase the resistance to the invasion of common phytopathogenic fungi and prolong its shelf life, improving the physical and chemical quality of the product. According to the results obtained by Saucedo-Pompa et al. [9], the concentration of antioxidant influences the speed of water loss, as it is directly proportional to the increase in the concentration of antioxidant.

At the beginning of the new millennium, a new era in the area of food and nutrition sciences has become increasingly intense: the area of food-medicine interaction increasingly recognized as the "functional foods" that accepts the role of food components, as essential nutrients for the maintenance of life and health and as non-nutritional compounds but that contribute to prevent or retard the Chronic diseases of the ripe age. Initially regarded as a passing curiosity, the idea of food formulation based on health benefits that its non-nutritional components could provide to the consumer, it has become an area of great interest today for large food companies [37], so the addition of natural antioxidant additives from the group of phenols such as ellagic acid and aloe vera to an edible cover, they represent a viable alternative for the development of nutraceutical roofs and enter in the field of functional foods [8]. In addition to offering a novel and comfortable presentation for the consumption of antioxidants by consumers and the VES improve the quality of shelf life in fruits [8] and avoid losses, by the attack of microorganisms, this due to the anti-fungal and antibacterial activity of the antioxidant additives [9].

11.6 PHENOLIC COMPOUNDS AS NATURAL ANTIOXIDANTS AND THEIR FUNCTIONAL ACTIVITY IN THE EDIBLE COATING

Phenols are also antioxidants and as such trap free radicals, preventing them from joining and damaging the molecules of deoxyribonucleic acid (DNA), a critical step in initiating carcinogenic processes, they also prevent lipid peroxidation, which, being free radicals can cause structural damage to normal cells, interfering with the transport of molecules through these membranes affecting cell growth and proliferation [38]. These phytonutrients, include a large group of compounds that have been subject to extensive research as preventive agents of diseases, which protect plants against oxidative damage and carry the same function in the human organism, whose main characteristic is its ability to block the action of specific enzymes that cause inflammation, but also modify the metabolic steps of prostaglandins and thus protect the agglomeration of platelets, according to data obtained from experimental studies, it seems that there are some possible mechanisms for the action of phenols, inhibiting the activation of carcinogens and therefore block the initiation of the process of Carcinogenesis [39]. Phenolic compounds come from barks, stems, leaves, flowers, organic acids present in fruits and phytoalexins produced in plants [40], such as caffeic, chlorogenic, p-coumaric, hydroxicinnamic, cinnamic, ferulic, and quinic acids that are present in plants which are used as spices, these acids present antimicrobial activity so they can retard the rot of fruits and vegetables, in fact, it has been shown that tannins and tannic acid also present antimicrobial activity [40]. Antimicrobial additives can inactivate essential enzymes, reacting with the cell membrane or altering the function of the genetic material, it has been observed that the pH and temperature affect the antimicrobial activity of these compounds [41].

The antioxidant compounds prevent the negative effects of free radicals on tissues and fats, reducing the risk of cardiac disturbances by avoiding the oxidation and cytotoxicity of LDL in vitro, decreasing the Aterogenicidad [42, 43]. Vitamins C, E, and β-carotene that prevent the oxidation of the LDL fraction of cholesterol, reduce the risk of coronary alterations, as well as possessing anti-cancer properties, whose protection measure consists of increasing the ingestion of fruits and vegetables, as well as foods containing antioxidant nutrients to protect from oxidation to LDL mentioned and thus avoid its oxidative modification and atherogenic formation [42].

Biomaterials such as Candelilla Wax have had a significant role in the coating of food, as it is a natural and biodegradable material, so it does not harm the environment besides being a natural safe-GRAS, generally Recognized as safe-recognized by the FDA, for its application in the food industry and can be used as an alternative in the elaboration of natural edible covers, reducing the use of synthetic polymers as derivatives of petroleum, which, when discarded, present a slow degradation in the environment compared with biomaterials, representing serious pollution problems. The edible wax covers of Candelilla, allow to control the respiration of the product, providing better permeability and texture, since it modifies the mechanical properties; Fulfilling the function of plasticizer. In addition, additives such as nutrients, dyes, antioxidants, antimicrobials can be added, which, as additives, make the edible cover a functional food.

At present, the realization of edible roofs is of lesser proportion compared with the elaboration of synthetic roofs, but the main advantage of edible covers made with biomaterial such as Candelilla wax, are easy to produce and quickly biodegradable, so it is a natural technique of conservation of fresh fruits.

It is important to consider the elaboration and commercialization of edible candelilla wax covers with the addition of antioxidants such as ellagic acid and aloe vera for the conservation of fresh fruit products in post-harvest, since it will allow to extend its shelf life avoiding microbial contamination, due to the antioxidant potential of ellagic acid and aloe vera, in addition to providing a benefit to the health of the consumer thanks to the anti-tumor characteristics, anticarcinogenic, and anti-cancer agents that present these antioxidants of natural origin.

ACKNOWLEDGMENTS

This work is part of the project "design of a high-performance process in the extraction of high-quality candelilla wax and formulation of products of final use from the prepared wax," financed by the project CONAFOR-CONACYT-S0002-2008-C01-91633.

KEYWORDS

- **biodegradable**
- **candelilla**
- **coatings**
- **edible**
- **wax**

REFERENCES

1. Tharanathan, R., (2003). Biodegradable Films and composite coatings: Past, present and future. *Critical Review in Food Science and Technology, 14*, 71–78.
2. Baldwin, E. A., Nisperos-Carriedo, M. O., & Baker, R. A., (1995). Use of edible coatings to preserve quality of lightly (and slightly) processed products. *Crit. Rev. Food Sci. Nut., 35*, 509–524.
3. Park, H. J., (1999). Development of advanced edible coatings for fruits, *Trends Food Sci. Technol., 10*, 254–260.
4. Miranda, M., (2003). Behavior of Chitosan films composed in an avocado storage model. *Revista de la Sociedad Química de México., 47*(4), 331–336.
5. Kester, J., & Fennema, O., (1986). Edible films and coatings, A review. *Food Technol., 40*, 47–59.
6. Multiceras (2009). http://www.multiceras.com.mx/pro-candelilla.htm.
7. FDA (U.S. Food and Drug Administration) (2009) http://www.fda.gov/.
8. Saucedo-Pompa, S., Jasso-Cantu, D., Ventura-Sobrevilla, J., Sáenz-Galindo, A., & Aguilar-Gonzales, C. N., (2007). Effect of candelilla wax with natural antioxidants on the shelf life quality of cut fresh fruits. *Journal of Food Quality, 30*, 823–836.
9. Saucedo-Pompa, S., Rojas-Molina, R., Aguilera-Carbo, A., Saenz-Galindo, A., De La Garza, H., Jasso-Cantú, D. et al., (2009). Edible film based on candelilla wax to improve the shelf life and quality of avocado. *Food Research International, 42*, 511–515.
10. Hagenmaier, R., (2005). A comparison of ethane, ethylene and CO_2 peel permeance for fruit with different coatings. *Postharvest Biology and Technology, 37*(1), 56–64.
11. Krochta, J., Baldwin, E., & Nisperos-Carriedo, M., (1994). Edible Coatings and Films to Improve Food Quality. *Technomic Publishing Company*, New York, p. 1344.
12. Bósquez-Molina, E., & Vernon-Carter, E. J., (2005). Effect of plasticizers and calcium on the water vapor permeability of mesquite gum based films and candelilla wax. *Revista Mexicana de Ingeniería Química, 4*, número 002, pp. 157–162.
13. Varoquax, P., Lecendre, I., Varoquax, M., & Souty, M., (1990). Changes in firmness of kiwifruit after slicing. (Perte de fermeté du fiwiaprés découpe) science des aliments. *10*, 127–139.
14. Cenamex (Ceras naturales mexicanas, S. A., de CV) (2009). http://www.cenamex.com.mx
15. Guilbert, S., Gontard, N., Morel, M. H., Chalier, P., Micard, V., & Redl, A., (2002). Formation and properties of wheat gluten films and coatings. In: Gennadios, A., (ed.), *Protein-Based Films and Coatings* (pp. 69–122). Boca Raton, FL: CRC Press.
16. Park, H. J., & Chinnan, M. S., (1993). Gas and water vapor barrier properties of edible films from protein and cellulosic materials. *Journal of Food Engineering, 25*, 497–507.
17. Pascat, B., (1986). Study of some factors affecting permeability. In: Mathlouthi, M., (ed.), *Food Packaging and Preservation, Theory and Practice*. Elsevier, Applied Science Pub. London. pp. 7–24.

18. Fennema, O., Donhowe, I. G., & Kester, J. J., (1994). Lipid type and location of the relative humidity gradient influence on the barrier properties of lipids to water vapor. *Journal of Food Engineering, 22*(1–4), pp. 225–239.
19. Bertuzzi, M. A., Armada, M., Gottifredi, J. C., Aparicio, A. R., & Jiménez, P., (2002). Study of water vapor permeability of edible films to coat food. Congreso nacional de ciencia y tecnología, buenos aires, Argentina, p. 220.
20. Bósquez-Molina, E., & Vernon, E. J., (2004). Effect of glycerol, sorbitol and calcium on the water vapor permeability of mesquite gum based films. XXV encuentro nacional AMIDIQ. resúmenes (ALI–21). 4–7 Mayo. Puerto Vallarta, México.
21. Koelsch, C., (1994). Edible water vapor barriers: Properties and promise. *Trends in Food Science and Technology, 5*, 76–81.
22. Ghaouth, E. L., Arul, J., & Ponnampalam, R., (1991). Use of chitosan coating to reduce water loss and mantain quality of cucumber and bell pepper fruits. *Journal of Food Processing and Preservation, 15*, 359–368.
23. Baldwin, E. A., Nisperos-Carriedo, M. O., Hagenmaier, R. D., & Baker, R. A., (1997). Using lipids in coatings for food products. *Food Technology, 51*(6), 56–61, 64.
24. Guilbert, S., & Biquet, B., (1986). Technology and application of edible protective films. En "*Food Packaging and Preservation.*" Editado por Mathlouthi, M. Ed. Elsevier. Londres.
25. Debeaufort, F., Quezada-Gallo, J. A., & Voilley, G., (1998). Edible films and coatings: Tomorrow´s packagings: A review. *Critical Rev. Food Sci. 38*(4), 299–313.
26. Soliva-Fortuny, R. C., & Martín-Belloso, O., (2001). Evaluation of zein films as modified atmosphere packaging for fresh broccoli. *J. Food Sci., 66*(8), 1108–1111.
27. Cuq, B., Gontard, N., & Guilbert, S., (1995). Edible films and coatings as active layers. En Rooney, M. L., (ed.), *Active Food Packaging* (pp. 111–135). London: Blackie Academic & Professional.
28. Fernandez, M., (2000). Review: Active packaging of foods. *Food Science and Technology International, 6*, 97–108.
29. Fira, (1997). Economic situation and prospects of avocado production in Mexico; Banco de México, S. A; División de Planeación, pp. 62–68.
30. Zhang, D., & Quantick, P., (1997). Effects of chitosan coatings on enzymatic browning and decay during postharvest storage of litchi (*Litchi chinensis* Sonn.) fruit. *Postharvest Biol. Technol., 12*, 195–202.
31. Buta, G. J., Moline, H. E., Spaulding, D. W., & Wang, C., (1999). Extending storage life of fresh-cut apples using natural products and their derivates. *Journal of Agricultural and Food Chemistry, 47*, 1–6.
32. Luo & Barbosa-Canovas, G. V., (1995). Inhibition of apple-slice browning by 4-hexylresorcinol. In: *Enzymatic Browning and its Prevention*. Washington, D.C., USA. America Chemical Society, 240–250.
33. Ruiz-Cruz, S., & Gonzáles-Aguilar, G. A., (2002). Effect of antioxidant agents in modified atmospheres in the quality of fresh pineapple slices. CIAD. Tesis de maestría. Hermosillo, Sonora, México.
34. Báez, R., Bringas, E., González, G., Mendoza, T., Ojeda, J., & Mercado, J., (2001). Post-harvest behavior of mango 'Tommy Atkins' treated with hot water and waxes. *Proc. Interamer. Soc. Trop. Hort.*, (USA), *44*, 39–43.

35. Gonzales-Aguilar, G. A., Monroy-Garcinia, I. N., Goycoolea-Valencia, F., Diaz-Cinco, M. E., & Ayala-Zavala, J. F., (2005). Edible covers of chitosan. An alternative to prevent microbial spoilage and preserve the quality of fresh cut papaya. *Proceedings of the Simposium "Nuevas Tecnologías de Conservación y Envasado de Frutas y Hortalizas*. Vegetales frescos cortados" La Habana, Cuba, 121–133.
36. Ghaouth, E. L., Ponnampalam, R., Castaigne, F., & Arul, J., (1992). Chitosan coating to extend the storage life of tomatoes. *Hort. Science, 27*, 1016–1018.
37. Best, D., (1997). All natural and nutraceutical. *Prepared Foods, 166*, 32–38.
38. So, F. V., Guthrie, N., Chambers, A. F., Moussa, M., & Carroll, K. K., (1996). Inhibition of human breast cancer cell proliferation and delay of mammary tumorigenesis by flavonoids and citrus juices. *Nutr. Cancer., 26,* 167–181.
39. Hertog, M. G., (1993). Dietary antioxidant flavonoids and risk of coronary heart disease: The zutphen elderly study. *Lancet, 342*, 1007–1011.
40. Beuchat, L. R., (2001). Control of food borne pathogens and spoilage microorganisms by naturally occurring antimicrobials. En: Wilson, C. L., Droby, S., (eds.), *Microbial Food Contamination.* CRS Press. London, UK. Chapter 11, 149–169.
41. Davison, P. M., (1996). Chemical preservatives and antimicrobial compounds. En: Doley, M. P., Beuchat, L. R., & Montville, T. J., (eds.), *Food Microbiology and Fronteirs* (pp. 520–566). Washington DC: ASM Press.
42. Seelert, K., (1992). Antioxidants in the prevention of atherosclerosis and coronary heart disease. *Internist Prax., 32*, 191–199.
43. Bello, J., (1997). Main clinical areas of application of functional foods or nutraceuticals. *Alimentación Equipos y Tecnología., 16*, 43–48.
44. International Trade Center (2009). www.intracen.org.
45. Villada, H., Acosta, H. A., & Velasco, R., (2007). Natural biopolymers used in biodegradable packaging. *Temas Agrarios., 12*(2), 5–13.
46. CONAFOR (National Forestry Commission). (2007). Technical report. Management of Commercial Forest Plantations. Zapopan, Jalisco, México.

INDEX

D

E

F

G

O

P

Printed in the United States
by Baker & Taylor Publisher Services